W9-BKL-938

THE ORIGINS
OF LIFE:

MOLECULES AND NATURAL SELECTION

THE ORIGINS OF LIFE:

MOLECULES AND NATURAL SELECTION

L. E. ORGEL

THE SALK INSTITUTE FOR BIOLOGICAL STUDIES
SAN DIEGO, CALIFORNIA

JOHN WILEY & SONS

New York   London   Sydney   Toronto

Copyright © 1973, by John Wiley & Sons, Inc.

All rights reserved. Published simultaneously in Canada.

No part of this book may be reproduced by any means, nor
transmitted, nor translated into a machine language without
the written permission of the publisher.

Library of Congress Cataloging in Publication Data

Orgel, Leslie E
  The origins of life.

  Bibliography: p.
    1. Life—Origin. 2. Chemical evolution.
I. Title. [DNLM: 1. Biogenesis. QH 325 0680
1973]
QH325.068   577   72-10534

ISBN 0-471-65692-5
ISBN 0-471-65693-3 (pbk)

Printed in the United States of America

10 9 8 7 6 5 4 3 2 1

PREFACE

This book is not written for professional biologists or chemists, but rather for college or advanced high school students and general readers who have a limited background in chemistry or biology. I have tried to show that studies of the origins of life have recently progressed to a point at which it is possible to propose plausible mechanisms for most of the steps in the evolution of living organisms from the inorganic constituents of the primitive earth. There are, of course, enormous gaps in our knowledge, but I believe that the origins of life can now be discussed fruitfully within the framework of modern chemistry and evolutionary biology. The extension of the discussion to life on other planets is straightforward, but necessarily very speculative.

I have adopted a number of procedures which would be out of place in a specialized monograph. Only the main current of thought on the origins of life is discussed and little mention is made of alternatives. I have not given references to the original literature, but instead have included a short bibliography that should enable the reader to trace ideas to their sources. I emphasize that I claim no priority for the ideas that I present; they are the work of many hands.

The sequence of chapters may occasion some surprise, for I have not always developed subjects systematically, as in a textbook. The logical structure of molecular biology, for example, is more easily grasped by the nonspecialist than the chemistry that underlies it. I have, therefore, inverted the traditional order and discussed the consequences of base-pairing, coding, etc. before discussing the chemistry involved. A detailed discussion of the optical activity of living systems has been relegated to an appendix late in the book, since I regard this topic as a special aspect of biochemical specificity.

Although the sub-title of this book is "Molecules and Natural Selection," I have little to say about natural selection and the origin of species. References to books dealing with that subject are given in the bibliography.

I am indebted to Dr. B. Chu, Dr. F. H. C. Crick, Dr. M. J. Dowler, Dr. J. R. Holum, Dr. H. A. Orgel, Mr. R. M. Orgel, Dr. C. Sagan, and Mr. R. Stileman for valuable comments on the original draft of my manuscript. I also wish to acknowledge my gratitude to the John Simon Guggenheim Memorial Foundation for a grant which greatly facilitated the completion of the work.

San Diego, California   **L. E. Orgel**

CONTENTS

PART
ONE

THE
NATURE
OF THE
PROBLEM

Historical Background

1

Introduction

The distinction between the living and the inanimate was one of the first to be drawn by man at the dawn of his cultural development. Most societies, whether primitive or advanced, have regarded the living world with special reverence. It is not surprising, therefore, that the origin of life has always been a subject of profound philosphical interest. If many of the opinions that were widely held even in comparatively recent times now seem naive, it is not because the people who held them were unintelligent. Although faced with problems that were beyond the technological capacity of their times, the early philosophers and biologists often gave remarkably farsighted answers to the fundamental questions about the nature and origin of life.

The last hundred years or so have seen an explosive expansion of our understanding of chemistry and biology. For the first time the problem of the origins of life is coming within range of the experimental and theoretical tools at our disposal. Later in this book we shall see how chemistry, biology, geology, and astronomy are beginning to throw exciting new light on the origins of life on earth. First, however, it is instructive to examine some of the opinions held in the past. This should at least prevent us from underestimating the problem; no doubt many of our own ideas will seem naive in the future.

Spontaneous Generation

Most scientists would agree that discussions of the origins of life should deal with processes that occurred long ago on the primitive earth and which led, after millions of years, to the emergence of living cells from lifeless starting materials. However, this point of view is a comparatively recent one. Most early cultures accepted that a god had created man and certain of the higher animals. They also thought that other organisms, such as insects and mice, were still being generated spontaneously from mud or decaying organic matter.

The theory of spontaneous generation was developed and systematized by the Greeks and received its most influential treatment in the writings of Aristotle (384–322 B.C.). For two thousand years from the time of Aristotle, educated men did not question the theory of spontaneous generation even in its simplest form. Aristotle's authority was not challenged in a serious way until the seventeenth century, and it was not until the middle of the nineteenth century that Pasteur finally showed that spontaneous generation does not occur. How could a theory that now seems obviously wrong and even absurb have survived so long?

First we must remember that the biological science of the Greeks was based very largely on casual observations, unaided by instruments such as the microscope. Just as important was the failure of the Greeks to appreciate that properly designed biological experiments are far more instructive than chance observations. To the casual observer it indeed appears obvious that young humans develop within their mothers, that birds emerge fully made from eggs, and that maggots form spontaneously in decaying meat. It requires a great deal more than casual observation to discover that the third conclusion is incorrect.

With all our modern background in biology and medicine it is hard to realize that until the seventeenth century intelligent, well-trained professional men, doctors and scientists, would have held all three conclusions to be equally self-evident. It required men of genius to question the theory of spontaneous generation, and it required the establishment of new standards for the design and interpretation of experiments finally to discredit it.

Aristotle, despite his unquestioned intellectual preeminence, was not always a careful observer. If he could state that men have more teeth than women, it is not too surprising that he thought that frogs could form from damp earth. It is a little more unexpected that van Helmont (1580–1644), one of the forerunners of scientific chemistry and biology, published a recipe for producing mice from soiled clothing and a little wheat. He was not too surprised when he found that these mice, despite their singular origin, mated successfully with normal animals.

It was an Italian biologist, Redi, who in a crucial series of experiments published in 1668, opened the attack on the theory of spontaneous generation. In retrospect Redi's experiments seem simple. It was well known that meat, when left to decay in the open, breeds maggots. Redi covered the meat with fine meshed muslin. He found that under these circumstances no maggots appeared. A careful examination of the muslin revealed the presence of eggs that were too large to pass through the mesh. After that it was not difficult to show conclusively that so long as insects and their eggs are kept away from decaying meat no maggots appear.

This demonstration did not deal a fatal blow to the theory of spontaneous generation, since, at that time, it was not clear that results obtained for a single species could be generalized to the whole living world. Redi himself never claimed to have disproved the theory of spontaneous gener-

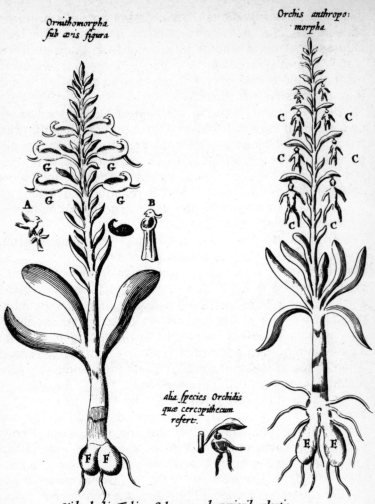

Ornithomorpha
fub avis figura

Orchis anthropo:
morpha

alia fpecies Orchidis
quæ cercopithecum
refert.

Vide de his Fabium Colonnam, de rariorib, plantis
Joh: B. Portam in phytognomia .

Figure 1.1. The end of the prescientific period.
Doves and anthropomorphic flowers growing on
orchids. An apparently normal dove is seen just
after detaching itself from the plant on the left.
What happened to the little men is not clear. This
figure is taken from the *Mundus Subterraneus* of
Athanasius Kircher, Amsterdam, 1665.

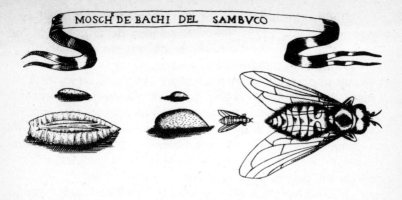

Figure 1.2. The beginning of controlled experimentation on generation. A typical life cycle worked out by Redi (for the Elder Fly). The figure is taken from *Esperienze Intorno alla Generazione degl'Insetti,* Florence, 1668. It first appeared only three years after the original of Figure 1.1.

ation; he maintained only that maggots did not arise spontaneously in meat. He and his contemporaries thought that the spontaneous generation of other insects probably did occur. Nonetheless, it is unlikely that the theory of spontaneous generation would have survived long had it not been for the work initiated by Anthony van Leeuwenhoek (1632–1723), one of the first microscopists.

During the latter part of the seventeenth century Leeuwenhoek published a series of detailed descriptions of the "animalcules" which he found to be very widely distributed, for example, in materials as different as rainwater and dung. Today, in place of "animalcule," we use the word microorganism. Leeuwenhoek seems to have discovered representatives of most of the major classes of microorganisms, including bacteria and yeasts. He showed that numerous "animalcules" could be seen whenever organic extracts were allowed to stand for long periods in contact with air and were then examined under the microscope.

Microorganisms are so small that they pass through muslin and the other materials which Redi had used to keep flies and their eggs away from meat. Thus it often happened that microorganisms penetrated barriers which, in Redi's time, were thought adequate for their exclusion. When this happened, new microorganisms seemed to appear spontaneously.

Since, at that time, many people still believed that insects arose spontaneously, it is hardly surprising that the spontaneous generation of these newly discovered microorganisms was often taken for granted. However, thanks to the work of Redi, this point of view was not held universally. Leeuwenhoek himself believed that microorganisms fell into his solutions from the air. In an attempt to settle the matter Louis Joblot, in 1718, carried out the first experiment of a type that was to be repeated with increasing attention to detail for the next century and a half.

Joblot boiled plant extracts for several minutes and then divided these sterile infusions into two portions. He left one of them in an open vessel and the other one in a vessel covered with parchment. Microorganisms appeared in the open vessel but not in the closed one. Next, Joblot removed the parchment covering and showed that this previously sterile solution soon developed its own population of microorganisms. Joblot concluded, correctly, that the microorganisms were not generated spontaneously.

Joblot's work did not convince his colleagues, and the conflict proved singularly difficult to resolve. In retrospect it is clear that many scientists of the eighteenth and nineteenth

The Nature of the Problem

centuries realized neither the difficulty of excluding bacteria completely from an organic solution, nor the resistance of certain organisms and their spores to heat sterilization.

John de Turbeville Needham (1713–1781) carried out a very extensive series of experiments from which he concluded that, even when all possible precautions are taken, microorganisms do appear spontaneously in previously sterilized solutions. Spallanzani carried out better experiments in 1765 and reached the opposite conclusion. He criticized Needham's experimental techniques; in particular he claimed that heat sterilization of the covered containers had been inadequate in Needham's experiments. Needham counterattacked by saying that Spallanzani had spoiled the broth and the air in his flask by heating them too intensely; microorganisms did not develop because no adequate nourishment remained. There seems little doubt that Spallanzani won this contest by means of brilliantly executed experiments, but the theory of spontaneous generation was so firmly entrenched that he failed to convince all of his contemporaries.

The technical problem of obtaining sterile solutions was aggravated by the realization that oxygen from the air has a profound effect on the decay of organic matter and is required for the growth of many types of microorganisms. The supporters of spontaneous generation quite legitimately insisted that ordinary "unspoiled" air be present along with the nutrients in any test of the theory. They claimed that air is spoiled by intense heating, so their opponents were forced to devise techniques for sterilizing a stream of air without exposing it to high temperatures. This difficult problem was finally solved by one of the greatest of all experimenters, Louis Pasteur.

The French Academy offered a prize for the most convincing experiments shedding light on the origins of living creatures. In 1862, Louis Pasteur won the prize for a series of experiments that finally discredited the theory of spontaneous generation. He analyzed the faults of previous experiments and then designed an experimental program of his own to eliminate them. His first step was to draw a stream of air through a pad of guncotton. The accumulated dust was

extracted with an organic solvent and the solid that remained was examined under the microscope. It proved to
contain microorganisms in great number and variety. Thus
the possibility that his solutions became contaminated by
airborne organisms could not be questioned.

Next Pasteur devised a number of ingenious procedures
for sterilizing a stream of cool air. Once he had solved this
problem he was able to show that no germs were formed in
previously boiled solutions. In one very beautiful series of
experiments, Pasteur partially filled a flask with broth and
then drew the narrow neck of the flask into an ∿. He next
boiled the contents of the flask long enough to sterlize them
and then allowed the flask and its contents to cool. Although
the flask was freely open to unheated air, its contents remained sterile for, as Pasteur had anticipated, all airborne
particles had been trapped on the curved surfaces of the ∿.
When the ∿-shaped neck was cut off, the broth soon began
to decompose, thus showing that it was still able to support
the growth of microorganisms once they got in. Since the air
in the vessel was never heated, Pasteur's opponents could
not claim that it was "spoiled."

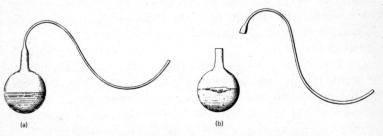

Figure 1.3. Louis Pasteur's flasks. The end of the
theory of spontaneous generation. The contents of
the flasks are boiled and then allowed to cool. (a)
No microorganisms appear because airborne particles cannot get past the ∿-shaped bend in the
tube. (b) Once the bend is cut off, the solution
develops a large population of microorganisms.
There is no question of "spoiling" the air or the
broth, since nothing is done to the body of the
flask or to its contents in going from (a) to (b).

These and many similar studies were of enormous importance for the development of bacteriology. They received great attention from the scientific community and were very widely believed to have shown that spontaneous generation is not a universal, contemporary process. They helped to raise the question of the origins of life in its modern form for the first time.

Recent Evolutionary Theories of the Origins of Life

Until the latter half of the nineteenth century it was generally believed in Western societies that God had created the higher animals, once and for all, "according to their kind." Pasteur's work showed that no living organisms, not even bacteria, come into existence except as the descendants of similar organisms. At earlier periods, the theory of special creation could perhaps have been extended to include microorganisms — God might have created them along with the higher animals but then decided not to mention them in Genesis. The development of Darwinian evolutionary theory soon made such explanations unacceptable.

Theories of the origins of living organisms have always been strongly influenced by contemporary religious and philosophical opinions. In the period immediately following Pasteur's work on spontaneous generation a very bitter public controversy was raging in England between orthodox Christians and Charles Darwin's more enthusiastic disciples. This gave a special flavor to subsequent discussions of the origins of life.

Darwin in his theory of evolution through natural selection proposed that species are not invariant, but change slowly with the passage of time. Darwin was not the first to advance a theory of evolution, but he was the first to support his theory with a mass of evidence sufficient to convince most men of science of his day. He and Alfred Russel Wallace were also the first to propose an acceptable explanation of evolution — natural selection or the survival of the fittest.

We shall return to this idea at greater length elsewhere.

11
Historical Background

Crudely, it was based on two generalizations. Firstly, in any species there is a good deal of variation from individual to individual. Secondly, children tend to resemble their parents more closely than they resemble members of the population chosen at random. Darwin argued that those individuals best adapted to their environment—the fittest—would survive to reproductive age more often than would less fit individuals. Hence the fittest individuals would, on the average, contribute more children to the next generation than would the less fit. Since children tend to resemble their parents, it follows that the children's generation would contain a higher proportion of fit individuals than had the parental generation. Thus, with the passage of time, the species would gradually change in the direction of better adaptation—the fittest would win out.

Darwin's theory differs radically from previous views on evolution, since it attributes change to the operation of chance. Variation within a species arises in a haphazard way; some changes are beneficial and others detrimental. Natural selection eliminates the detrimental variations and preserves those that are beneficial. This is a very different point of view from that put forward in the evolutionary theories current before the time of Darwin and Wallace. According to some of these theories evolution is willed; a species improves because parents pass on to their descendants "information" that they have acquired from the environment during their lifetime. In other pre-Darwinian theories of evolution, the environment is supposed to act directly on the members of a species and to bring about inheritable changes in them.

Most of Darwin's supporters were prepared to generalize his theory and to assume that living organisms could have arisen from the inorganic world by an evolutionary process. Many Christians found this hard to accept, since it implied that the evolution of primitive organisms from inorganic matter and of man from primitive organisms could have occurred without the intervention of any supernatural creative act. Thus the problem of the origins of life became a central issue in the wide-ranging controversy between religion and biology. More than a hundred years later the State

Board of Education in California takes essentially the same position as Darwin's opponents: biology textbooks used in schools in California are required to give equal weight to evolutionary theories of the origins of life and to the story of creation recounted in the Bible.

Darwin himself was cautious in his published writings, but there is no doubt that he believed that life could have originated spontaneously on the primitive earth. The following quotation from one of his letters makes this quite clear.

> It is often said that all conditions for the first production of a living organism are present, which would ever have been present. But if (and oh, what a big if) we could conceive in some warm little pond, with all sorts of ammonia and phosphoric salts, light, heat, electricity, etc., present, that a protein compound was chemically formed ready to undergo still more complex changes, at the present day such matter would be instantly devoured or absorbed, which would not have been the case before living creatures were formed.

While Darwin and many other nineteenth and early twentieth century scientists believed that life might have evolved from inorganic materials on the primitive earth, they appreciated that organic chemistry had not yet advanced to a point at which an experimental attack on the problem of chemical evolution could prove useful. The reaction of the Swedish chemist, Svente Arrhenius, to the fall of the doctrine of spontaneous generation is typical of that of another group that included many distinguished chemists and physicists. He believed that, since life was no longer appearing afresh on the earth, it must have been introduced from another planet. Arrhenius, in a series of scientific papers and popular books, developed a theory called Panspermia according to which organisms, aided by the pressure of radiation, could have made the long journey across space from another solar system.

Neither this idea, nor the somewhat related notion that life was carried to the earth on a meteorite, is ridiculous, although both are open to serious objections. However, even if correct they do not solve the problem of the origins of life but merely transfer the problem to another planet. To overcome this criticism, Arrhenius and many of his supporters argued that life must be eternal. In this way they abolished the problem by decree—since life is eternal the question of its origin does not arise.

Today this view seems mystical. Presumably, each planet has evolved from inorganic dust and simple gases floating freely in space. Hence, it seems that the spontaneous creation of living organisms on any planet by other than evolutionary processes would require supernatural intervention. Arrhenius' theory shifts the site of the origin of life to another planet, but does not provide a mechanism for the origin of life.

In summary, little important work on the origins of life was carried out in the period immediately following the revolutionary publications of Pasteur and Darwin. Experimental work was largely restricted to fruitless attempts to prove that microorganisms can be produced in special environments, for example, in organic material subjected to the influence of radium. Although some scientists believed that life had evolved from inorganic matter long ago, they also believed, correctly, that experimental approaches to the problem were premature. A large and influential group did not believe that life could have evolved spontaneously and accepted either explicity or implicity the need for supernatural intervention.

It may not be an accident that the next major step forward was taken in a society that had deliberately adopted a materialistic philosophy and was actively antireligious. In 1923, in Russia, A. I. Oparin suggested that the atmosphere of the earth, long ago, was very different from the present atmosphere. In particular, it did not contain oxygen but rather hydrogen and other reducing compounds, such as methane and ammonia. Oparin proposed that the organic chemicals on which life depends formed spontaneously in such an atmosphere, under the influence of sunlight, lightning, and the high temperatures existing in volcanos. A similar suggestion was made independently by another materialist, J. B. S. Haldane, in England.

These two proposals influenced most of the authors whose speculations about the origins of life were published

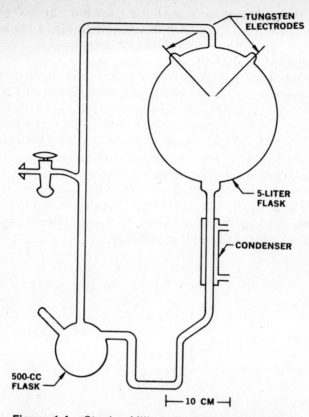

Figure 1.4. Stanley Miller's flask. The beginning of prebiotic chemistry. (See Chapter 8 for a full description of these experiments.) Reproduced with permission from *J. Am. Chem. Soc.*, **77,** 2352 (1955).

during the 1930s and 1940s. It is surprising that experimental confirmation was delayed for so long, since the theoretical notions had such wide currency. It was not until 1953 that Stanley Miller, working with Harold Urey in Chicago, demonstrated that important biochemicals are indeed formed in surprisingly large amounts when an electric discharge is passed through an atmosphere of the kind proposed by Oparin and Haldane.

These experiments were the beginning of the most recent phase in the study of the origins of life. They led to a detailed consideration of the conditions that must have existed at the surface of the primitive earth and to attempts to reconstruct in the laboratory the chemistry that would have occurred under those conditions. Describing this work is one of the objectives of the present book.

There has, however, been a second major influence on our ideas about the origins of life. In 1953, the Watson-Crick structure of DNA was announced and since then molecular biology has developed at an accelerating pace. Recent discoveries concerning the structure of the genetic system and the way in which it operates have sharpened enormously our understanding of what needs to be explained by a successful theory of chemical evolution.

The Fossil Record

2

Geological Dating

The fossil record is our only source of information about the antiquity of life on earth. Yet, until recently, it was not possible to determine the ages of fossils. Within the last twenty years all this has changed. A new technique called radioisotope dating has been introduced which allows geologists to determine the ages of fossils with considerable accuracy. Reliable estimates of the ages of some very early forms of life can now be given.

Before the introduction of radioisotope dating, geologists were sometimes able to determine the relative ages of different geological strata by making use of two very simple principles. When layers of rock have been formed by sedimentation, that is by deposition at the bottom of oceans or lakes, the lower layers must clearly be older than the ones that lie above them. This principle does not apply to layers

17

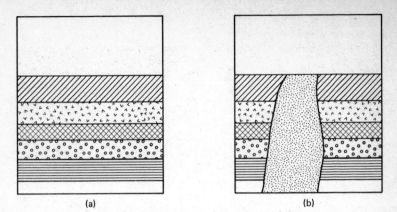

Figure 2.1. Ordering of rocks according to their age: (a) the strata get older as one goes downward through a sedimentary column; (b) the intrusion must be younger than any of the sedimentary layers that it cuts.

of igneous rock, that is to layers of rock which have been formed by the cooling of molten material extruded from the interior of the earth. Here, however, the second simple principle was sometimes applicable. If a layer of one type of rock cuts through a stratum of quite a different kind, then the layer that is unbroken must be the more recent.

Using this kind of information it was often possible to order the strata in a locality according to their relative ages, even when the strata had been disturbed by earth movements and mountain building. It is important to notice, however, that these methods do not provide any evidence about the absolute ages of the strata, nor do they help to determine the relative ages of rocks found in widely separated places.

Relative ages could sometimes be determined by an examination of the fossil record. The fossils found in sedimentary rocks represent the creatures that lived when the rocks were being deposited. Sometimes fossils are nothing more than molds of the harder parts of organisms. Sediment often fills the interior of clam shells, for example, and then gradually hardens. The shells later dissolve away leaving

The Nature of the Problem

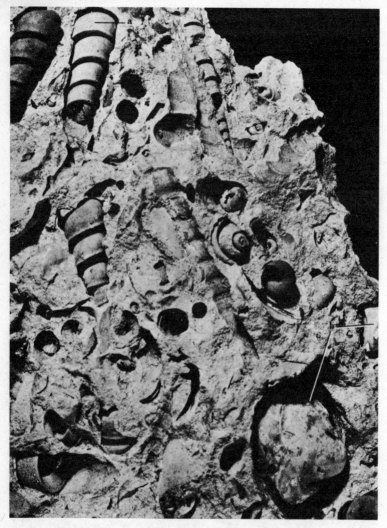

Figure 2.2. Internal moulds in Portland limestone. The moulds are gastropod and bivalve shells. (Reproduced with permission from *The Elements of Palaeontology* by R. M. Black, Cambridge University Press, New York, 1970.)

behind internal molds. External molds develop in a similar way; some represent the exterior of the complete animal while others are restricted to as small a detail as a footprint or a worm's burrow.

The most familiar types of fossils are those formed by replacement. Most shells are made of calcium carbonate. In the course of time slightly soluble silicates contained in ground water often replace the more soluble calcium carbonate. The calcium phosphate which makes up the bones of vertebrates can be replaced by silicate in a similar way. Remarkably faithful replicas of the harder parts of long-dead

Figure 2.3. A well-preserved fossil *Echinoderm* (sea-urchin). (Reproduced with permission from *The Elements of Palaeontology* by R. M. Black, Cambridge University Press, New York, 1970.)

The Nature of the Problem

creatures are formed by this process. Since silicate fossils are much more durable than shells or bones, they outlast the remnants of the original animals.

To make use of fossils in determining the relative ages of rocks, we must assume that if two rocks contain the same kinds of fossil they are equally old. This amounts to supposing that each species persisted through roughly the same period of earth history in all of its habitats. Originally this assumption was questioned, but once it had been accepted it enabled geologists to establish relations between strata in widely different localities. Thus if fossils of the same species were found in strata in Europe and North America, the strata were deduced to be equally old.

By applying these simple ideas with great ingenuity and industry, geologists and palaeontologists were able to piece together an extraordinarily detailed picture of the sequence of animals that have inhabited the earth. Their evidence was one of the foundations on which Darwin based his theory of evolution. The problem of absolute ages, however, remained to be solved; work on the order in which species have succeeded each other did not provide information about the time periods involved. Early estimates of the ages of fossils were based on indirect evidence and were very far from correct.

The next major step forward in geology depended on radioiostope dating, a technique made possible by advances in the chemistry and physics of radioactive elements. Radioactive elements are, of course, unstable. They emit high-energy radiation and are thereby transformed into new elements. The products of radioactive decay usually have chemical properties completely different from those of the materials from which they are formed. The gas, argon, for example, is formed by the decay of the metal, potassium.

The study of radioactivity is complicated, mainly because most chemical elements occur as several different isotopes. By this we mean that several kinds of atom, each having a different mass (weight) occur together in samples of the chemical element. Carbon, for example, occurs as four different isotopes, ^{11}C, ^{12}C, ^{13}C, ^{14}C, with masses 11, 12, 13, and 14, respectively. One of these isotopes, ^{12}C, is much more abundant than the others. Although all forms of carbon have the same chemical properties, the rare isotope ^{14}C is radioactive, while the abundant isotope ^{12}C is stable.

From our point of view, the most important characteristic of a radioactive isotope is its half-life. Each radioactive atom has a fixed probability of decaying per unit time. This probability is unaffected by the environment of the atom and by the time that the atom has already survived. Atoms get older, but they do not age. We define the half-life of an isotope as the length of time needed to give an atom a one-in-two chance of decaying. Equivalently, it can be defined as the time it takes for half of the atoms in a large sample to decay. Half-lives vary from fractions of a second to billions of years.

The idea on which radioisotope dating is based is easily understood. Let us oversimplify and suppose that the isotope of potassium, ^{40}K, decays to an argon isotope, ^{40}Ar, with a half-life of 1.26 billion years. Then, if we found that a rock specimen contained equal numbers of ^{40}K and ^{40}Ar atoms and if we could be sure from the nature of the rock that all of the ^{40}Ar must have been formed by the decay of ^{40}K, we could deduce that the rock was one half-life old, that is 1.26 billion years old. If we found three ^{40}Ar atoms for each ^{40}K atom we would know that three quarters of the potassium had decayed and hence that the rock was two half-lives old, that is 2.52 billion years old, and so on. To determine the age of the rock all we would need to know is the relative abundances of ^{40}K and ^{40}Ar.

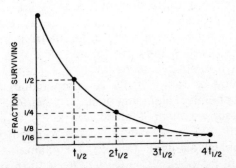

Figure 2.4. The decay of a radioactive element as a function of its half-life.

The Nature of the Problem

Real elements do not behave as simply as we have supposed. ^{40}K decays to ^{40}Ar, with a half-life of 1.26 billion years, but it also forms a calcium isotope ^{40}Ca. Difficulties of this kind are not serious; they complicate the arithmetic but they do not limit the applicability of the method. Major difficulties are usually encountered when the radioactive element or its decay products could have been lost from the specimen. Soluble compounds are often washed out when water percolates through a rock, and volatile compounds escape when the rock is heated. Consequently, porous rocks that have been exposed to water, or rocks that have been heated, are often unsuitable for dating in this way.

Three important methods of radioisotope dating have been applied to old sediments. They are based on the decay of uranium to lead, rubidium to strontium, and potassium to argon, respectively. In every case the abundance of the radioactive element and of its decay products can be measured accurately. By choosing rock specimens with care and by employing more than one method of dating whenever this is feasible, it has been possible to establish the ages of many of the oldest geological formations.

The Classical Fossil Record

Geology and palaeontology were well-developed sciences long before absolute dates could be assigned to fossil remains. This led to the development of unsystematic nomenclature, which is baffling, except to the geologist. In Figure 2.5 we give the geological eras and periods together with the ages which have been assigned to them by radioisotope dating. The terms cenozoic, mesozoic, and palaeozoic are self-explanatory, at least to Greeks and a few others. They refer to supposedly recent, middle, and ancient periods in the development of life. The names of the periods have varying derivations; many are named after localities where their formations are particularly well-represented.

The oldest fossils of largish animals date from the early Cambrian, about 580 million years ago. These creatures were marine invertebrates; by the end of the Cambrian period about 500 million years ago, sponges, jellyfish, starfish,

THE GEOLOGIC TIME SCALE

ERA	PERIOD	EPOCH	MILLIONS OF YEARS AGO (APPROX.)	DURATION IN MILLIONS OF YEARS (APPROX.)		RELATIVE DURATIONS OF MAJOR GEOLOGICAL INTERVALS
CENOZOIC	QUARTERNARY	RECENT / PLEISTOCENE	0–1	1		CENOZOIC
						MESOZOIC
	TERTIARY	PLIOCENE	1–13	13		
		MIOCENE	13–25	12		PALEOZOIC
		OLIGOCENE	25–36	11		
		EOCENE	36–58	6		
		PALEOCENE	58–63			
MESOZOIC	CRETACEOUS		63–135	72		
	JURASSIC		135–181	46		
	TRIASSIC		181–230	49		
PALEOZOIC	PERMIAN		230–280	50		
	PENNSYLVANIAN		280–310	30		
	MISSISSIPPIAN		310–345	35		
	DEVONIAN		345–405	60		PRECAMBRIAN
	SILURIAN		405–425	20		
	ORDOVICIAN		425–500	75		
	CAMBRIAN		500–600	100		
PRECAMBRIAN	UPPER / MIDDLE / LOWER	Although many local subdivisions are recognized, no world-wide system has been evolved. The Precambrian lasted for at least 2½ billion years. Oldest dated rocks are at least 2,700 million, possibly 3,300 million, years old.				

Figure 2.5. The geological time scale showing the standard divisions based on the fossil record and the lengths of time obtained from dating rocks. William Lee Stokes, *Essentials of Earth History*, 2nd ed., © 1966. Reprinted by permission of Prentice-Hall, Inc., Englewood Cliffs, N. J.

and marine worms were all common. However, the dominant animals were the trilobites; some species were as small as a pinhead and others as much as eighteen inches long. The trilobites in time became extinct, but many of the other Cambrian organisms are clearly related to animals that can still be seen today in tidepools.

24
The Nature of the Problem

Fish were the first vertebrates to appear. Fragments of "bony" coverings of fish-like creatures are found in deposits that are more than 420 million years old, but fossils of fish much more than 400 million years old are uncommon. It seems that a remarkably rapid evolution of a great variety of fishes occurred in the Devonian period about 380 million years ago. The new species included some true fishes, such as the sharks, and also scaly fishes with lungs and limb-like fins. These latter were the ancestors of the amphibians.

The first amphibians were quite small, but by 300 million years ago giant salamander-like creatures had already taken to the land. The earliest land plants evolved from seaweed or

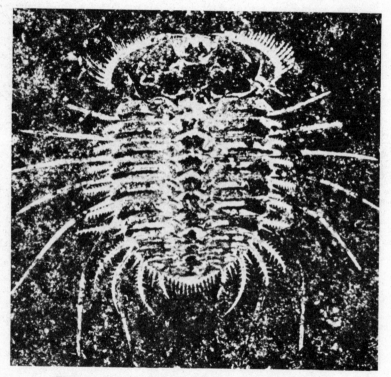

Figure 2.6 A spinose trilobite from Czechoslovakia (×4). (Reproduced by permission of Professor H. B. Whittington of the Sedgwick Museum, Cambridge.)

The Fossil Record

other algae about 400 million years ago; 100 million years later the land was covered with ferns and primitive trees. It was in this environment, perhaps 280 million years ago, that the first reptiles appeared and then gradually displaced the amphibians. The earliest remains of modern insects date from about the same time.

The period from 180 to 60 million years ago was the age of the reptiles. These included the giant dinosaurs and flying pterosaurus as well as more familiar lizards, turtles, and crocodiles. The first mammals evolved from reptiles about 180 million years ago, but they did not become dominant for more than another 100 million years. A quite different evolutionary branch led from the reptiles to the modern birds.

The mammals began to take over the land a little less than 60 million years ago. All of the modern groups of mammals appeared within the next 25 million years. Thirty-five million years ago, primitive horses, pigs, rodents, and monkeys were already established.

The mammals continued to evolve and diversify. Palaeontological evidence suggests that man's ancestors diverged from other groups of apes about 20 million years ago, but biochemical evidence suggests a much shorter time since the separation, perhaps only five million years. Very recently the ice ages, which ended 10–15,000 years ago, caused the extinction of many plants and of the giant mammals. They also resulted in great changes in the distribution of surviving species. Since then the environment has been comparatively stable, except where it has been transformed by agriculture or other human activities.

Fossil Microorganisms

The creatures of the Cambrian period were the oldest to leave fossils only because they were the first to form hard shells. These fossils represent complicated, multicellular organisms which must have had a long evolutionary history. Fortunately, fossils of microorganisms from earlier times have survived. The examination of Pre-Cambrian microfossils began early in this century, but has only recently begun to produce detailed information about these earliest forms of life. In part, this rapid increase in our knowledge is due to the use of the electron microscope.

One of the most completely studied groups of Pre-Cambrian microfossils is that from the Gunflint formation located on the north shore of Lake Michigan in Canada.

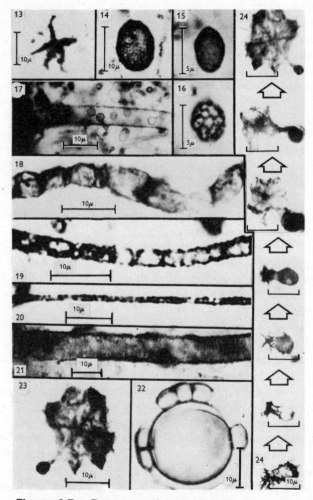

Figure 2.7. Representative organisms from the Gunflint Chert. (Reproduced with permission from J. W. Schopf, *Biol. Rev.,* **45,** 319, (1970).)

The Fossil Record

Although the rocks in which these fossils are found are about 1.8 billion years old, many of the fossils resemble very closely species of blue-green algae which are still living today.

Fossil remains from earlier periods are few in number and less well-preserved. This is because extremely old, unmodified sedimentary rocks are rare. In the course of time most of the oldest sediments have spent periods in the interior of the earth where they have been heated to such an extent that no fossils have survived. Very few sedimentary rocks more than 3 billion years old are known. The simplicity of the structures of the oldest microfossils is another source of difficulty, since it is sometimes impossible to distinguish fossils from inorganic artifacts.

These problems have led to occasional controversy amongst experts. Nonetheless, it is generally accepted that microfossils resembling contemporary blue-green algae are present in rocks three billion years old or somewhat older. The microfossils in the Fig Tree series from Africa (3.1 billion years old) are perhaps the best preserved. It is also clear that the earliest Pre-Cambrian fossils are the simplest, and that microorganisms of increasing complexity evolved throughout the later Pre-cambrian. Thus the classical fossil record despite its enormous variety, corresponds only to the most recent 20% of a continuous record covering more than 3 billion years.

We have emphasized that Pre-Cambrian fossils and modern algae look very much alike. How do we know that the biochemistry of Pre-Cambrian organisms was similar to modern biochemistry? A new and rapidly developing branch of palaeontoloy which deals with chemical fossils attempts to answer this question. Chemical fossils are organic substances that are found in fossil-bearing rocks and that derive from the original organisms. Several important biochemicals have been found in association with fossil algae in very old sediments. Unfortunately much of the evidence is still very controversial, so while on the whole it supports the idea that biochemistry has changed little since the early Pre-Cambrian, it would be wise to reserve judgment until more evidence is available.

The Nature of the Problem

grain
boundary

Figure 2.8. A simple bacterium-like particle about 3 billion years old from the Pre-Cambrian of South Africa (Fig Tree Series). Notice how difficult it is to be sure that a particle of this kind is really the fossil of a living organism. (Reproduced with permission from E. S. Barghoorn and J. W. Schopf, *Science*, **152**, 758 (1966).)

The evidence that we have discussed shows that it is very probable that single-celled organisms resembling modern blue-green algae were already present on the earth 3.0 to 3.5 billion years ago. It does not in any way prove that this was the time when they first appeared. The earth is about 4.5 billion years old and we know virtually nothing about the first part of its history. The most that we can say is that life appeared at some time during the first billion or so years after the formation of the earth.

In order to develop a feeling for the vast times that have passed since the beginning of life on the earth, it is helpful to think in terms of the number of elapsed generations. There have been about 100 human generations since the flowering of Greek civilization and 500–1,000 since the last ice age. About 50,000 generations have gone by since the origin of man. By contrast, much more than 10 million generations separate us from the earliest mammals. The lineage of modern insects is even longer, running to hundreds of millions of generations.

Living organisms similar to bacteria and algae have existed on the earth for three billion years or more. We do not know how long the most primitive organisms took to divide; since modern bacteria under ideal conditions take about 20 minutes to reproduce, let us assume an average generation-time of two or three hours. Modern bacteria are then calculated to be the product of some 10,000,000,000,000 generations of evolution. Even if our assumptions are incorrect, this estimate could hardly be off by more than a factor of ten. It is astonishing that, as we shall see, much of the chemical organization of cells is so stable that it seems to have remained unchanged throughout this enormous number of generations.

Questions Posed by the Fossil Record

The most obvious questions raised by the fossil record concern the nature of the evolutionary processes that led to the formation of new species. How did complex organisms evolve from the very simple algae and bacteria that lived more than three billion years ago? What conditions favored the evolution of the mammals? Which creatures are the direct ancestors of man? The fossil record also raises a second set of questions concerned with the origins of life itself. What happened in the first billion years of earth history that led to the appearance of organisms similar to modern bacteria? Was there one origin of life, or were there many?

Most elementary introductions to biology deal with the origin of species. They show that the theory of natural selection is adequate to explain the evolution of complex organisms from simpler ones. This subject is not of primary interest to us, since it deals with a very late stage in the history of life. We shall concentrate on the evolution of the most primitive cells from the inorganic constituents of the primitive earth. Natural selection will have to be discussed extensively, but for the most part only in this specialized context. The reader interested in the more conventional applications of the theory should consult the books cited in the bibliography for an account of natural selection and the origin of species.

Living organisms represent the ultimate in miniaturization; the machinery of life is constructed on the atomic scale. The simplest living things are incredibly complicated; they are so much more complicated than anything in the nonliving world that even the largest modern industrial complexes seem relatively simple when compared with the smallest living cells. If we have interpreted the fossil record correctly, equally minute and comparably complicated organisms had evolved on the earth three billion years ago. It is the enormous gap that must be bridged between the most complicated inorganic objects and these simplest living organisms that provides most of the intellectual challenge of the problem of the origins of life.

Before we can begin to discuss the crucial transition from relatively simple inorganic systems to living organisms, we shall need to describe in some detail the end product of the origins of life—the most primitive living organisms. We know very little directly about the chemistry of the organisms that lived on the earth three billion years ago, but we may infer a great deal from the behavior of modern organisms. The next two chapters describe the growth and reproduction of the simplest modern organisms, and in Chapter 5 this information is used to infer as much as possible about the behavior of the corresponding Pre-Cambrian organisms.

Molecular Biology

3

Cells and Proteins

Clearly we cannot hope to understand how life began without knowing something about the way in which living things work. In this chapter we shall survey some modern ideas on the structure and function of the simplest living organisms.

Cells are the basic unit of life. Bacteria, the least complicated of living organisms, consist of a single cell; higher animals may contain thousands of billions of cells. Whereas bacteria cells are small and simple, animal cells are often large and complicated. Nerve cells, for example, are sometimes many feet long. (See Figure 3.1).

We shall see in Chapter 5 that all modern species are descended from bacteria or algae that lived on the earth billions of years ago. In considering the origins of life, therefore, it is natural to take the bacterial cell as the prototype of

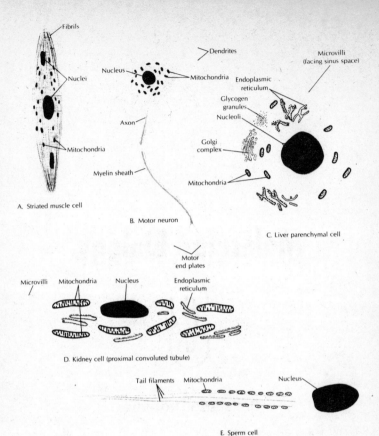

Figure 3.1. Some typical animal cells. The diagrams are not drawn to scale. They demonstrate the great differences between different kinds of cells in complex organisms. Each cell type is adapted to its own specific function. (Reproduced with permission from A. G. Loewy and P. Siekevitz, *Cell Structure and Function,* 2nd ed., Holt, Rinehart and Winston, Inc., New York, New York, 1969).

all living cells. Asking how life began is equivalent to asking how the first unicellular organisms evolved. We cannot formulate that question correctly until we know something about modern unicellular organisms.

The chemical processes that enable bacteria to grow

and divide are so complicated that they cannot be described in the way that one describes the operation of a simple machine. Instead we must follow a more roundabout route, first describing the broad strategy of bacterial growth and later coming back to fill in some of the details.

The bacterium, *Escherichia coli* (*E. coli*), has been studied more extensively than any other microorganism. Since it is a relatively simple unicellular organism, it may conveniently be used to illustrate most aspects of bacterial growth. *E. coli* cells are rods $2-3 \times 10^{-4}$ cm long and 5×10^{-5} cm in diameter; more than 1,000,000,000,000 cells could be packed into a volume of one cubic cm.

The bacterium is enclosed by a rigid cell wall and by a cell membrane which is situated just within the wall. The cell wall serves to protect the membrane from damage. Cells grown in the presence of penicillin lack cell walls and are consequently very fragile. The role of the membrane is in part passive; it acts as a barrier to keep essential molecules inside the cell and to prevent the entrance of harmful molecules. The cell membrane also plays a more active part in bacterial growth, for it contains a series of "pumps" which select useful molecules from the outside and concentrate them within the cell. Thus the membrane maintains the correct balance between the contents of the cell and the external environment.

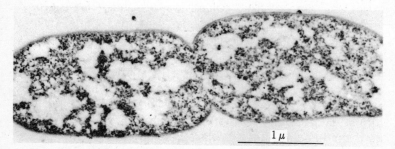

1 μ

Figure 3.2. An electron micrograph of cell division in *E. coli*. (Reproduced with permission from S. F. Conti and M. E. Gettner, *J. Bacteriol.* **83**, 544 (1962).)

The main business of the cell is carried on within the membrane. Here, a series of some hundreds of coordinated chemical reactions take place. The sole purpose of all this activity is to use nutrients in the environment to make more *E. coli.* This task is accomplished with impressive efficiency; under ideal conditions a complete cycle of replication takes only twenty minutes. Thus one *E. coli* cell could produce $2^{72}(10^{21}-10^{22})$ descendants in a single day.

The individual chemical reactions that take place within the cell are quite similar to those carried out in the laboratory. However, in the cell they all take place in aqueous solution at room temperature. The organic chemist can make use of special solvents and carry out reactions at very high or very low temperatures. Even so he cannot yet match the synthetic ability of the cell.

The remarkable transformations achieved within living cells are made possible by a set of molecules called enzymes. Enzymes are catalysts; that is, they are molecules which speed up chemical reactions without themselves being changed in the process. They are able to function repeatedly. So, one enzyme molecule, if given sufficient time, can transform many times its own mass of material. The substances on which enzymes act are called substrates. Each minute, a typical enzyme transforms a few thousand substrate molecules into products.

Enzymes are proteins. The proteins constitute a family of substances whose molecules are very large and are made up of "building blocks" called amino acids. Thus the proteins are polymers ("poly-" many; "-meros", parts) having amino acids as their monomers ("mono-" single; "-meros", parts). We can compare a protein molecule to a long word made up of a number of letters (the monomers) some of which may be used many times. Just 20 kinds of amino acids are used in the construction of proteins.* (A few more are formed occasionally by the modification of the twenty standard amino acids after they have been incorporated into proteins.)

---

* For a discussion of the optical activity of proteins and nucleic acids see the appendix to Chapter 10.

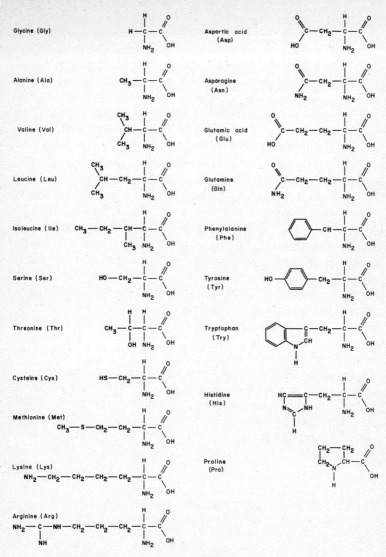

Figure 3.3. The twenty naturally occurring amino acids. Standard abbreviations are given in parentheses.

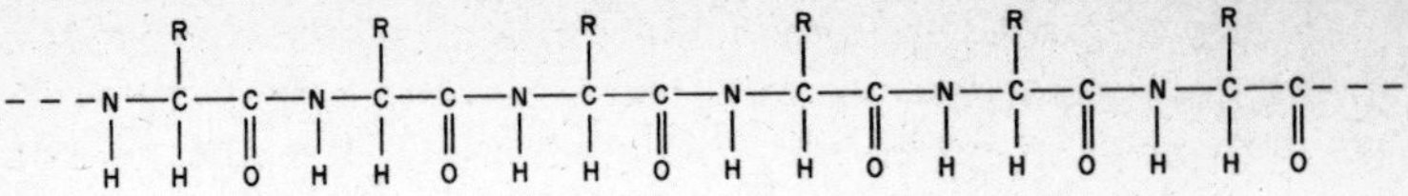

Figure 3.4. The standard method of joining amino acids to form peptides. R can be any one of the twenty side groups shown in Figure 3.3. From *Molecular Biology: Genes and the Chemical Control of Living Cells* by J. M. Barry, Prentice-Hall, Inc., Englewood Cliffs, N. J., pp. 3 and 4.

The uniqueness of a word or sentence depends on the choice of letters and the sequence in which they are put together. In a similar way, the uniqueness of the protein depends on the choice of amino acids and the sequence in which they are joined. Just as a single alteration can change the meaning of a sentence, so, changing a single amino acid may alter the properties of a protein. Consequently, the health and well-being of an organism depends on having the correct proteins made *exactly* right every time. Getting the sequences exactly right is largely the job of another family of polymers in a cell, the nucleic acids—the chemicals of heredity. We shall discuss the nucleic acids in the next section.

Proteins are the most versatile of biological molecules, for all chemical reactions in cells are catalyzed by proteins. In addition, proteins are essential for the operation of the pumps in the cell membrane. Most *E. coli* strains have fibers called flagella attached to their membranes. These fibers, which enable the organisms to "swim," are made of protein.

```
 Lys Val Phe Gly Arg Cys Glu Leu Ala Ala Ala Met Lys
 Gly Leu Ser Tyr Gly Arg Tyr Asn Asp Leu Gly His Arg
 Asn Trp Val Cys Ala Ala Lys Phe Glu Ser Asn Thr Gln
 Gly Tyr Asp Thr Ser Gly Asp Thr Asn Arg Asn Thr Ala
 Ilu Leu Gln Ilu Asn Ser Arg Trp Trp Cys Asn Asp Gly
 Cys Pro Ilu Asn Cys Leu Asn Arg Ser Gly Pro Thr Arg
 Ser Ala Leu Leu Ser Ser Asp Ilu Thr Ala Ser Val Asn
 Asn Met Gly Asp Gly Asp Ser Val Ilu Lys Lys Ala Cys
 Ala Trp Val Trp Arg Asn Arg Cys Lys Gly Thr Asp Val
             Len Arg Cys Gly Arg Ilu Trp Ala Gln
```

Figure 3.5. Amino acid sequence of chicken lysozyme. Note the complexity of a typical protein. For abbreviations see Figure 3.3.

In higher animals the contractile parts of muscle are made up of protein and so are important parts of nerve cells. All dynamic processes in cells are mediated by proteins. (The *energy* needed to bring about these processes is provided by sugars, fats, starches, and oils in the diet.)

Each enzyme performs a unique function in the cell. Several hundred enzymes carry out routine production steps; many more can be brought into action under special circumstances, for example, when an essential substance in the environment begins to run out and an alternative source of it must be found. In many cases the activity of an enzyme is regulated by the intracellular environment. In this way the rates of the various processes going on in the cell are adjusted to make growth and maintenance as efficient as possible. As in a modern assembly plant, raw materials are transformed into finished products by the coordination of large numbers of simple operations.

Nucleic Acids and Protein Synthesis

The analogy between a cell and a factory is a useful one, but it fails to draw attention to the most remarkable property of living systems, their ability to reproduce. Since a cell can divide to give two identical daughter cells, there must be a mechanism for duplicating the cellular machinery, that is, for making new protein molecules with exactly the same sequences as those originally present in the parental cell. At first sight, it might seem that the easiest way to do this would be to copy each individual protein molecule, letter by letter. In fact, the synthesis of the correct protein molecules is achieved by a quite different method.

Proteins are not duplicated directly; instead, the information needed to specify a protein sequence is encoded in the sequence of a totally different polymeric molecule, a nucleic acid. The nucleic acids are important because they are the only molecules in the cell that can, with the help of appropriate enzymes, replicate directly. As we shall see, the direct replication of nucleic acids, that is the process in which a nucleic acid molecule directs the synthesis of a new molecule with exactly the same sequence, is the essential process that guarantees that daughter cells resemble their parents so closely. In higher organisms, also, nucleic–acid replication is responsible for all aspects of biological inheritance.

There are two kinds of nucleic acids in every cell— deoxyribonucleic acid (DNA) and ribonucleic acid (RNA). Each time a cell divides, the genetic nucleic acid, DNA, is duplicated accurately and one copy is passed on to each daughter cell. The DNA carries all of the information needed to direct the synthesis of new cellular proteins. However, DNA takes no direct part in protein synthesis. Instead, DNA functions as a master-copy from which RNA sub-copies are made and these RNA sub-copies are the ones that carry the genetic information to the protein-synthesizing system. It is as though the DNA was a benevolent dictator sending "doubles" to represent itself in situations where it might otherwise be damaged.

The nucleic acids never act as ordinary catalysts. If the proteins are thought of as the machinery of the cell, the genetic nucleic acids must be thought of as the blueprints. There is a division of effort within the cell; proteins are responsible for most cellular activity while nucleic acids make possible the storage and transmission of genetic information.

DNA is a polymer made up of four types of small molecules called deoxynucleotides. To understand how the nucleic acids fulfil their dual function in the cell, we must look rather closely at their chemical structures. Each deoxynucleotide consists of a deoxyribose phosphate molecule attached to one of four bases. It is the nature of the base that distinguishes one deoxynucleotide from another. The deoxynucleotides that make up DNA are usually designated as A,

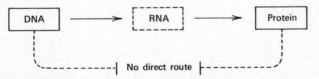

Figure 3.6. The role of RNA as an intermediate in protein synthesis. DNA does not take a direct part in protein synthesis, although it ultimately controls the nature of the protein synthesized.

The Nature of the Problem

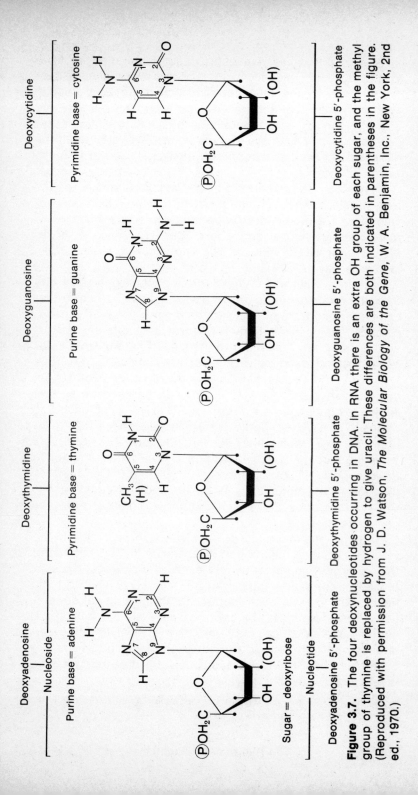

Figure 3.7. The four deoxynucleotides occurring in DNA. In RNA there is an extra OH group of each sugar, and the methyl group of thymine is replaced by hydrogen to give uracil. These differences are both indicated in parentheses in the figure. (Reproduced with permission from J. D. Watson, *The Molecular Biology of the Gene*, W. A. Benjamin, Inc., New York, 2nd ed., 1970.)

T, G, and C,* which are abbreviations for adenylic acid, thymidylic acid, guanylic acid, and cytidylic acid, respectively. Since the deoxynucleotides that make up a nucleic acid are joined together in a regular way, we can represent any DNA molecule as a sequence of these four letters, for example, TCATTGTC. We include an arrow to show that the direction of reading is important, just as it is in an English word or sentence.

The four components of RNA are very similar to the deoxynucleotides that made up DNA; they are called ribonucleotides. The sugar deoxyribose is replaced by a closely related sugar, ribose, and one of the four bases, T, is replaced by uracil (U)* (Figure 3.7). The chains are joined up in the same way in RNA and DNA.

The replication of DNA and RNA or the copying of one from the other depends on a very remarkable structure that can be formed from polynucleotide chains. A pair of polynucleotide chains fits together to form a beautifully regular double-helix, but only if the sequences of ribonucleotides or deoxynucleotides in the two chains are complementary. (See Figure 3.9) It is necessary that, after the direction of one chain has been reversed, the two chains can be lined up so that each A is opposite to a T (A opposite U in RNA) and each G is opposite to a C (see Figure 3.10). The sequences ACTAAGC and GCT-TAGT match after the second sequence is reversed, while the sequences ACTAAGC and GGTTACT fail to match in the second and sixth positions. The rules, $A = T$ and $G = C$, that govern the matching of bases are known as the Watson-Crick pairing rules. In the next chapter we shall see how they come about.

The genetic material, DNA, is always stored in cells in the form of a double-helix. When it replicates the two strands of DNA behave independently and each strand directs the synthesis of a complementary strand. In this way two new double helices are formed, each identical with the original double-helix. (See Figure 3.11) Mistakes are very rare, because if the wrong base does go into one of the newly synthesized strands, the regularity of the corresponding double-helix is disturbed. Then DNA replication stops until the incorrect base is removed.

* Adenine (A) and guanine (G) are purine bases. Thymine (T), uracil (U), and cytosine (C) are pyrimidine bases.

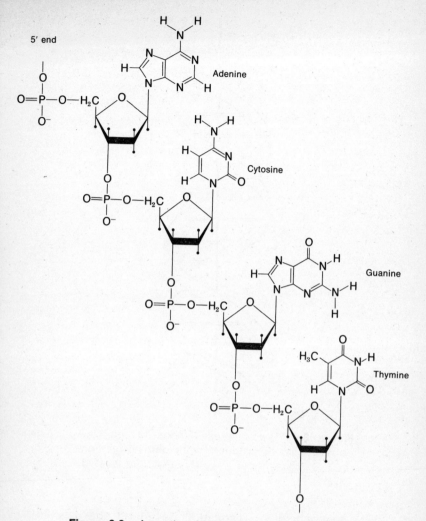

5' end

Adenine

Cytosine

Guanine

Thymine

Figure 3.8. A section from a DNA chain showing the sequence ACGT. (Reproduced with permission from J. D. Watson, *The Molecular Biology of the Gene,* W. A. Benjamin, Inc., New York, 2nd ed., 1970.))

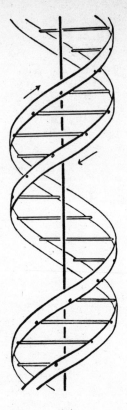

(a)

Figure 3.9. Two representations of the DNA double-helix. (Reproduced with permission from (a) J. D. Watson and F. H. C. Crick, *Nature,* **171,** 737(1953.))

The mechanism of replication described above is the only one that is important in cells. In certain viruses, however, RNA is used instead of DNA as a genetic molecule. The RNA replicates directly and also functions directly in protein synthesis. It should also be noted that many viruses contain single-stranded nucleic acids. They replicate in almost the same way as individual strands of double-helical DNA. (See Figure 3.12)

No other polymeric molecules are known which are able

The Nature of the Problem

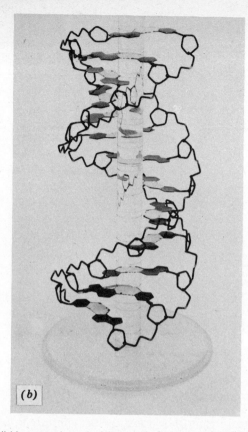

(b) Is reproduced with permission from F. H. C. Crick,
in *Molecular Basis of Life,* W. H. Freeman, San Fran-
cisco, 1968.

to combine together to give compact two-chain complementary structures. The Watson-Crick pairing rules, which form
the basis of the molecular theory of heredity, are, as far as
we know, unique. Since it seems unlikely that any replicating
structure could be made up from amino acids, we can understand why the information needed to determine the
sequence of a protein must be encoded in a non-protein
polymer, such as a nucleic acid; there is no structural basis
for the direct letter-by-letter replication of proteins.

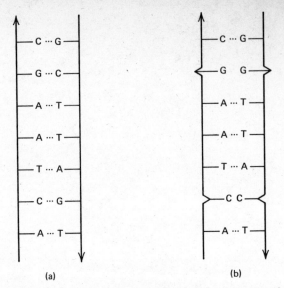

Figure 3.10. (a) Two complementary strands of DNA; (b) two DNA strands that fail to match because of GG and CC mispairing.

How Proteins are Made

We have seen that DNA does not act directly in protein synthesis. Instead, one of its strands functions as a template which lines up the components of RNA according to the Watson-Crick pairing rules and thus leads to the synthesis of strands of RNA complementary to the active DNA strand. These newly synthesized RNA molecules are known as messenger RNA, since they act as intermediaries carrying the genetic information stored in the DNA sequence to the protein-synthesizing apparatus. The series of operations by means of which the sequence of nucleotides in messenger RNA chains determines the sequence of amino acids in protein molecules is called translation. Clearly, translation must be quite complicated since messages written in the four-letter alphabet of the nucleic acids are used to specify the twenty letters which appear in the protein translations.

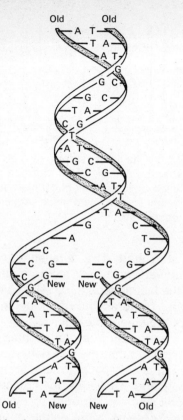

Figure 3.11. A diagrammatic representation of the replication of DNA. The details of the process are not understood at the time of writing. (Reproduced with permission from J. D. Watson, *Molecular Biology of the Gene*, W. A. Benjamin, Inc., New York, 2nd ed., 1970.)

It is perhaps surprising that the principle involved in translation is simple and familiar. The letters of the nucleic acid message are always read three at a time from a fixed starting point. (See Figure 3.13) Each triplet of nucleotides corresponds either to an amino acid or to a signal to stop translation. Since there are 64 ($4 \times 4 \times 4$) different triplets and only 20 amino acids, many of the amino acids are represented by two or more triplets. In addition three triplets corre-

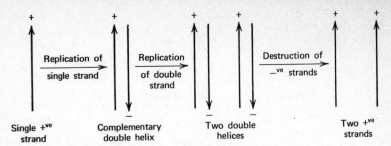

Figure 3.12. The simplest form of replication of a single strand RNA or DNA. Most single strand viruses used a more economical method in which several positive strands are made sequentially on a single negative strand.

spond to the stop sign. The detailed rules for translating nucleic acid sequences into protein sequences have been worked out in the last decade or so. The solution is known as the genetic code. The cracking of the genetic code is one of the triumphs of molecular biology.

Translation is much more complicated than replication. So, it is not surprising that the translation apparatus is more complicated than the replication apparatus. More than a hundred protein and nucleic acid molecules take part in translation. Replication is brought about by one, or at the most a very few, proteins. More details about translation are given in Chapter 4.

We are now in a position to see how the parts of the genetic system fit together. Two sets of molecules, proteins and nucleic acids, cooperate to allow cells to grow and divide. Nucleic acids store and transmit genetic information; proteins do almost everything else in the cell (except provide

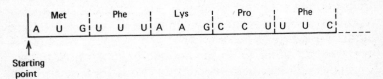

Figure 3.13. Translation of messenger RNA into protein. See Figure 3.14 for the code.

The Nature of the Problem

		U		C		A		G		
U	U	UUU UUC	Phe	UCU UCC	Ser	UAU UAC	Tyr	UGU UGC	Cys	U C
		UUA UUG	Leu	UCA UCG		UAA UAG	OCHRE AMBER	UGA UGG	UMBER Tryp	A G
F I R S T L E T T E R	C	CUU CUC CUA CUG	Leu	CCU CCC CCA CCG	Pro	CAU CAC	His	CGU CGC CGA CGG	Arg	U C A G
						CAA CAG	GluN			
	A	AUU AUC AUA	Ileu	ACU ACC ACA ACG	Thr	AAU AAC	AspN	AGU AGC	Ser	U C A G
		AUG	Met			AAA AAG	Lys	AGA AGG	Arg	
	G	GUU GUC GUA GUG	Val	GCU GCC GCA GCG	Ala	GAU GAC	Asp	GGU GGC GGA GGG	Gly	U C A G
						GAA GAG	Glu			T H I R D L E T T E R

Figure 3.14. The genetic code. There are three stop signs that signal the end of a polypeptide chain—UAA, UAG, UGA. All polypeptides are initiated with the AUG codon for methionine. The methionine is often removed from the end of the chain by a special enzyme.

energy). Nucleic acids could not replicate or direct protein-synthesis without the help of preformed proteins; no proteins could be synthesized without the information stored in preformed nucleic acids. (See Figures 3.15 and 3.16). One of our major problems in later chapters is to see how such a "hen and egg" relation could have developed.

Cellular Function

A more detailed account of the way in which cells function must now be presented. For simplicity, bacteria will be discussed. Most bacteria derive energy and material for growth from nutrients in their environment. The membrane contains proteins called permeases which select the molecules the cell needs from outside and pump them into the cell. It is

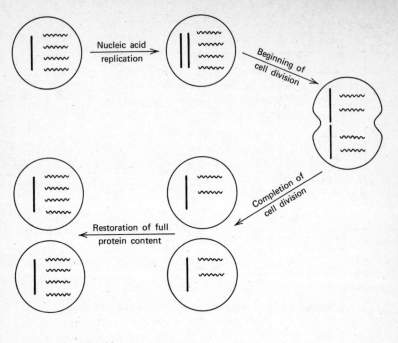

————— = Nucleic acid

〰〰〰〰 = Protein

Figure 3.15. The logic of cell division. All details have been oversimplified.

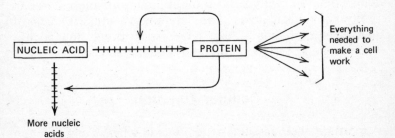

Figure 3.16. ———→ Enzyme functions of proteins; ⧺⧺⧺⧺▷ Information-transferring functions of nucleic acids in which the sequence of a preformed nucleic acid determines the sequence of a newly formed nucleic acid or protein.

quite unclear how many different permeases *E. coli* can make; the total number could be as high as one hundred.

E. coli cells are quite flexible in their nutritional demands. Many different organic compounds are suitable as sources of food, for example, amino acids, glycerol, and glucose. If no suitable nutrient is directly available in the environment, *E. coli* cells can often produce extracellular enzymes which convert complex substances in the environment, for example, starch, into simple nutrients. Furthermore, most *E. Coli* strains are able to detect distant sources of food and will, under appropriate conditions, move towards them.

Once inside the cell, the raw materials from the environment must be converted into all of the small molecules required for growth. Clearly, very many steps are involved in converting simple nutrients, such as glucose and ammonia, into amino acids, nucleotides, fats, etc. Several hundred enzymes that help to synthesize important biochemicals have already been identified; the total number could exceed one thousand.

Another major cellular activity is the synthesis of polymers (e.g., proteins, nucleic acids, starches, etc.). We have already seen that the synthesis of proteins requires more than a hundred macromolecules while the replication of DNA is much simpler and requires very few. These are not the only proteins involved in polymer synthesis. Many scores of enzymes are required, for example, to synthesize additional components of cell walls and cell membranes.

There is another group of enzymes within the cell that searches out damaged RNA and protein molecules and degrades them so that their components can be used again. Since no cell can reproduce if its genetic material has been destroyed, damaged DNA molecules have to be dealt with in quite a different way; they are repaired rather than degraded. In recent years a group of enzymes has been discovered, each of which is able to repair a particular kind of defect in damaged DNA. Without these repair activities, cells would often be unable to divide to give viable daughter cells.

Energy is expended in almost every one of the steps which we have described so far. Different cells derive their energy from different sources, but all of them use a single chemical, ATP, as the principal intermediate in the utilization of energy. Photosynthetic bacteria use the energy of sunlight directly to make ATP, but most other bacteria obtain the energy needed to make ATP by breaking down nutrients, such as sugars, into simpler molecules. *E. coli,* for example, can derive energy by converting glucose to carbon dioxide and water.

Processes of this kind have been described as "controlled burning," but this phrase is somewhat deceptive. A typical sugar, such as glucose, will burn in air to give carbon dioxide and water. Energy is released in the process as heat (and a little light). Although a heat engine could be run on glucose, the living cell has a different and better way of making use of this and related substances. The cell employs about twenty enzymes to dismantle glucose, step by step. The whole series of reactions is achieved mildly at room temperature, and about half the available energy is channelled directly into the synthesis of ATP, without ever appearing as heat. Thus, although heat engines and cells sometimes bring about the same overall chemical changes, the oxidation of nutrients by air to carbon dioxide and water, they do it in different ways.

Cells produce energy in many other ways. In the absence of air, for example, they often break down glucose to lactic acid. Processes like this, which do not require air, are called fermentations; they produce much less ATP for each nutrient molecule consumed than do oxidations. Other kinds

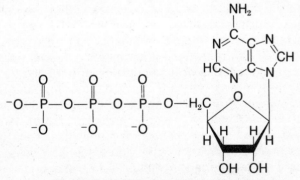

Figure 3.17. The structure of ATP.

of cells are able to derive all of their energy from inorganic sources.

Despite the diversity of energy sources used by different kinds of cells, the major energy-utilizing processes are strikingly similar in all cells. ATP is used, often indirectly, in the synthesis of nucleic acids and proteins, in many syntheses of small molecules, in the repair of damaged nucleic acids, in pumping, and so on. Of the major groups of activities which we have discussed so far, only the degradative reactions can continue indefinitely after the supply of ATP is cut off. Cells degenerate without the use of energy, but new synthesis is absolutely dependent on ATP. In Chapter 4 the structure of ATP and the way in which the energy stored in ATP is made to do useful work are described in greater detail.

So far we have dealt with proteins which catalyze chemical reactions, transport molecules across membranes, or act directly in some other way on the contents of the cell or on substances in its environment. A distinct class of proteins is involved in controlling and coordinating all these activities. The proteins of this class are able to use information about the internal and external environments of the cell to control the activity of preformed enzymes or the synthesis of new enzymes.

It is worth looking at two typical examples. Suppose that cells that have been growing on glucose are transferred to a medium containing a different sugar, say lactose. At first no enzymes are present that can utilize lactose. So, growth stops. However, within a short time new enzymes are synthesized that can act on lactose. Then growth resumes. This adaptation to a new medium is achieved through the action of a "control" protein which recognizes that lactose is the best available nutrient in the environment and then causes the enzymes that act on lactose to be synthesized.

Suppose next, that after cells have been growing for some time on glucose as the sole nutrient, a complete mixture of amino acids is introduced into the environment. Clearly it would be uneconomical for the cell to continue making amino acids when they are freely available. Amino acids are, therefore, pumped in from the outside and the enzymes making amino acids are very quickly shut off. Later a further economy is achieved by stopping the synthesis of all the enzymes required to make amino acids, since these enzymes are not needed as long as the external supply of amino acids lasts.

It should now be clear that at any given time the enzymes functioning within a cell constitute only a fraction of those which the cell is able to make. To deal with a changing environment, cells must have a reserve capacity to make many proteins that are only occasionally required.

What is the inside of a cell like? We know that water is essential for life, but how much of the interior of a cell is left over for water once proteins, nucleic acids, and assorted small molecules have been fitted in? In general, cells are filled with a concentrated jelly-like solution of protein. In *E. coli,* for example, about 80% of the cell content is water and most of the rest is protein. Other cells are similar, although the detailed composition of the intracellular medium varies from organism to organism.

Viruses

The simplest cells are self-contained reproducing units; they will grow and divide in environments that contain nothing but suitable nutrients and a few inorganic salts. Viruses have much in common with cells, but they differ from them in one important and clear-cut way. Viruses cannot reproduce without the help of the macromolecular machinery inside the cells of their hosts.

The simplest viruses contain a short nucleic acid which codes for a very small number of proteins. Every virus is able to code for the one or more proteins which form a protective coat around its nucleic acid, and most viruses also produce an enzyme to replicate their RNA or DNA. Small viruses are completely dependent on their hosts for a supply of the components needed to build proteins and nucleic acids. More important, they are dependent on the protein-synthetic apparatus of their hosts; they do not carry the complex protein-synthetic apparatus needed to express the information encoded in their nucleic acid. Larger viruses can carry out some additional functions for themselves, but they are still dependent on the synthetic capacity of their hosts.

It has sometimes been suggested that the first organisms were viruses rather than cells. This suggestion is based on the mistaken notion that viruses are autonomous self-replicating units. In fact, since viruses cannot replicate without the help of a host cell, the evolution of viruses could not have occurred before the appearance of cellular forms of life. From this point of view, at least, viruses should not be regarded as living organisms.

Molecular Genetics

The genetic material of *E. coli* is physically continuous. It forms a double-helical DNA molecule containing about 4,000,000 base pairs. However, the genetic material is functionally discontinuous in the sense that it controls the synthesis of some thousands of discrete kinds of protein molecules. The sequence of DNA responsible for the synthesis of a single protein is called a gene. A typical protein might contain 200 amino acids; the corresponding gene would contain 600 base pairs. Genes are arranged in linear sequence along the DNA molecule, separated by an elaborate scheme of punctuation marks. The punctuation marks are themselves particular nucleotide sequences, but special ones that are recognized by the translation apparatus to mark the beginnings or endings of standard messages.

Early in the development of microbial genetics, the relation between the genetic material and the functional elements of cells was summed up aptly in the phrase "one gene—one enzyme." Since then we have learned a great deal about the detailed operation of the genetic system, but this generalization remains a valuable one. We analyze the relation between the functional materials and the genetic materials of any organism in terms of the proteins and their corresponding DNA sequences—the enzymes and the genes that specify them.

Our knowledge of genes rests on very strong physical evidence. Genes are no longer mysterious abstractions in-

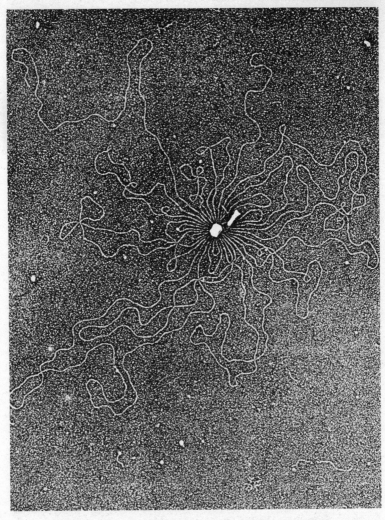

Figure 3.18. The DNA released from a damaged bacteriophage. DNA in *E. coli* is about ten times longer. (Reproduced with permission from A. K. Kleinschmidt, *Biochim. Biophys. Acta,* **61,** 857(1962).)

voked to explain complicated facts about inheritance in plants and animals. Genes can be seen in the electron microscope and their lengths can be measured. The order of hundreds of genes on *E. coli* DNA is known with certainty. A very simple gene has been synthesized in the laboratory. There are still some difficult problems in bacterial genetics, but the nature of the gene and of its relation to cellular proteins is understood, at least in broad outline. The organization of the genetic material in higher organisms, that is, the structure of chromosomes, is understood less well.

The description of the genetic system that we have given so far contains an important supposition: DNA is replicated without error. Recent studies with *E. coli* show that in fact errors do occur with a frequency of perhaps one in 10^7; that is, on the average, one error occurs for every 10,000,000 bases copied. Since *E. coli* DNA contains about 4,000,000 letters, there is almost a one-in-two chance of a mistake occurring in each round of replication. In the course of many millions of generations, the accumulation of mistakes has important consequences.

Mistakes in DNA replication are called mutations. The most common mutations are substitutions of one nucleotide for another or additions and deletions of a single nucleotide. More complicated and extensive errors occur, but less frequently. The usual consequence of a mutation is the production of a protein with a modified amino acid sequence. Three typical mutations, a substitution, an insertion and a deletion, are illustrated in Figure 3.19.

Strangely, perhaps, it is errors in the synthesis of DNA which permit bacterial evolution. If no errors occurred, each microorganism would produce descendants with an unchanged DNA sequence, and hence, with an unchanged

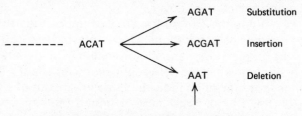
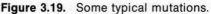

Figure 3.19. Some typical mutations.

genetic potential; families derived from a single ancestor would remain exactly like that ancestor for generation after generation. The occurrence of occasional errors in the replication of DNA is the only source of variety in those bacteria for which sexual processes are unimportant. Of course, in higher animals the reassortment of genetic material associated with sexual reproduction provides a second, and often more important, source of variation.

Most mutations are harmful or, at best, neutral. A mutated form of an enzyme is usually inactive or less efficient than the original form; a mutated control protein usually switches on or off at the wrong times. Occasionally, mutations are advantageous if they produce modified proteins better able to carry out an old function, or if they produce a protein able to perform a new and useful function.

It is interesting to compare mutations with errors made in setting the type for a book. Most random errors lead to nonsense or, what is worse, to a deceptive new sense. Rarely, a new and quite unanticipated meaning may result from an error. The change from "the comforts of immortality" to "the comforts of immorality" in a religious tract is a striking example of a deletion.

One should note that, even within a single cell, the effects of a mutation may be out of all proportion to the physical change that occurs in the DNA. A mutation involving a single base, if it leads to the production of an inactive variant of an essential enzyme, can kill a whole cell. There is no relation between the physical extent of the change in the DNA and the importance of the change for the cell. Returning to the analogy with printing, it is as though some mutations deleted the words "do not" from an essential section of an instruction manual, while others merely shortened the recommendations on the dust cover.

Many mutations can be transmitted unchanged through a large number of generations. Once an error has occurred, it may be copied unchanged for millions of generations. Thus, a single mistake in DNA replication can affect the performance, not only of an individual bacterium, but of a whole species. These arguments apply in much the same way to higher organisms. Sickle-cell anemia, for example, is a not

58
The Nature of the Problem

uncommon disease in which the oxygen-carrying capacity of the blood is impaired. All carriers of the disease are descended from one or a very few ancestors in whom the mutation occurred. The degree of amplification involved in an instance like this is hard to match in the nonbiological world. A single event, at the molecular level, has affected the lives of a very large number of individuals.

We can now see why mutations, almost all of which are individually harmful, permit a species to become better adapted to its environment. Organisms that inherit the disadvantageous mutations are sooner or later eliminated by the competition of their fitter neighbors. On the other hand, organisms that inherit one of the very few beneficial mutations are likely to grow at the expense of their neighbors and will probably displace them. This is part of the molecular interpretation of Darwin's theory of natural selection. The biological world is dominated by the lineages that invent useful adaptations; failures are soon forgotten.

Topics in Biochemistry

4

Introduction

This chapter brings together a number of unrelated topics in chemistry and molecular biology that will be useful in later chapters. Only the section on base-pairing is essential for an understanding of the general argument of the book, but the other sections should prove useful to most readers. Those who have difficulty with chemistry are advised to read quickly through all the material except that on base-pairing. They may find it useful to return to other sections when they are referred to later in the text.

As we all know, chemistry deals with the properties of atoms and molecules. An atom consists of a central positively charged nucleus surrounded by a cloud of negatively charged electrons just sufficient in number to cancel out the positive charge of the nucleus. The carbon atom, for example, consists of a nucleus of charge $+6$ surrounded by a cloud of 6 electrons.

When atoms are brought together, their electronic clouds become deformed. In many cases the deformation is of such a type that the atoms attract each other strongly and combine to form molecules. However, when the nuclei get too close, the repulsive force between their positive charges always overcomes any attractive force due to the deformation of the electron clouds. Thus, the nuclei in molecules do not approach each other indefinitely closely but settle down at a distance known as the interatomic distance (but literally the internuclear distance). The distance between neighboring nuclei in a molecule is characteristically about 10^{-8} cm or 1 Å. A molecular chain 1 cm long contains about 10^8 atoms.

For many purposes molecules can be treated as solid objects, each with a shape and size determined by the rules of valency and stereochemistry. The chemical rules of va-

Figure 4.1. Molecular models of some typical molecules: (a) water (H_2O), (b) benzene (C_6H_6), (c) adenine ($C_7N_7H_7$).

Figure 4.2. Tetrahedral arrangement of four groups around carbon in methanol (methyl alcohol).

lency specify the numbers of bonds that can be formed by each type of atom. Additional rules describe the stereochemistry of atoms, that is, the way in which the bonds formed by an atom are arranged in space. A carbon atom, for example, in one of its characteristic environments, is surrounded by four groups at the apexes of a regular tetrahedron. Recently far subtler rules have been discovered that throw additional light on the spatial arrangements of atoms in complex molecules, such as proteins.

We do not need to understand these rules in detail. For the purposes of this book it is sufficient to realize that in almost all molecules the different groups of atoms are arranged in a well-defined structure. Each group of atoms has certain properties that, to a first approximation, are independent of its position in the molecule, particularly if the different active groups are not too close together. A knowledge of the properties and positions of the groups of atoms in a molecule is all that is needed to explain many molecular properties. We shall first illustrate this point by describing some properties of an important group of molecules called surface-active agents.

Surface-Active Agents and Colloids

It is possible to classify the atoms or groups of atoms that make up organic molecules into those that tend to enhance and those that tend to diminish the solubility of a substance in water (Table 4.1). These groups are called hydrophilic and hydrophobic, because they tend, respectively, to seek

63

Table 4.1

Hydrophobic Groups	Hydrophilic Groups
—CH$_3$	—OH
—CH$_2$—	—NH$_2$
$\overset{\mid}{\underset{}{—CH—}}$	$—C\overset{\diagup O}{\diagdown OH}$
—S—CH$_3$	—SO$_3$H

out or to avoid water. Most sugars, for example, are very soluble in water because their molecules contain many hydrophilic OH groups, while most petroleum products are insoluble because their molecules are almost exclusively made up of hydrophobic CH, CH$_2$, and CH$_3$ groups.

When substances containing many hydrophilic groups dissolve in water, they form true solutions in which the dissolved molecules are isolated from each other by a layer of water. Substances containing hydrophobic groups are usually insoluble in water because water molecules tend to stay together and squeeze out hydrophobic intruders, while hydrophobic molecules tend to remain together and avoid water.* The reasons why water molecules and hydrophobic molecules do not mix can be understood in terms of physical and chemical theories, but the explanations are too difficult to be given here.

Molecules that have a hydrophilic head and a hydrophobic tail have quite exceptional solubility properties, because the heads tend to dissolve in water while the tails have a tendency to stay in contact with each other. Such molecules are called surface-active agents. They include soaps and detergents.

Surface-active agents often accumulate at boundaries, for example, at the interface between oil and water. In this way both ends of the molecules of the surface-active agent can be accommodated in favorable environments: the head

* Many molecules are insoluble for different reasons.

remains in the aqueous phase while the tail dissolves in the oil. The importance of detergents as cleansing and emulsifying agents depends on their ability to bridge the interface between a hydrophobic substance and water.

When surface-active agents are shaken with water, they often aggregate to form associations of molecules in which the hydrophobic tails are packed together inside while the hydrophilic heads stick out. In this way the hydrophobic parts of the molecule are able to remain together while allowing the hydrophilic parts to establish contact with the aqueous environment.

If a sufficient number of molecules associate, we refer to the aggregate as a colloidal particle, and we refer to the mixture of colloidal particles with water as a colloidal dispersion (solution) or simply as a colloid. A colloid is a mixture which resembles a solution in that it contains more or less uniform particles dispersed in a solvent in such a way that they do not settle out. However, the solute is present in aggregates much larger than the single organic molecules present in a simple solution. Colloidal dispersions have a number of special properties different from those of simple solutions, for example, they scatter light in a characteristic way. The particles in a colloidal dispersion are smaller than those present in suspensions or precipitates.

Surface-active agents are only one of a large class of molecules that form colloidal dispersions. The Russian scientist A. I. Oparin has emphasized the importance of certain

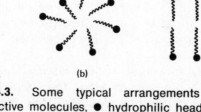

(a)   (b)   (c)

Figure 4.3. Some typical arrangements of surface-active molecules, ● hydrophilic head, ∿ hydrophobic tail: (a) bridge at an oil-water boundary, (b) a colloidal droplet, (c) a bilayer membrane.

colloidal dispersions, which he refers to as coacervates, in the origin of life. Both high molecular-weight polymers and simple surface-active agents could have contributed to the formation of these aggregates.

Another important structure formed by surface-active agents is the bimolecular leaflet. Here sheets of molecules are brought together with their hydrophobic tails inside and the hydrophilic heads in contact with water. Artificial bilayers of this kind form membranes that are impermeable to electrically charged atoms and molecules. Similar bilayers are thought to be important components of biological membranes.

The Structure of Biological Polymers

(a) Base-Pairing and the Structure of Nucleic Acids. The nucleic acids are the molecules responsible for all forms of inheritance. Without them, the evolution of living organisms similar to those that we know would have been impossible. Almost all of the functions of nucleic acids depend on base-pairing; one cannot appreciate the problem of the origins of life without understanding this one aspect of structural chemistry.

For nucleic acids to perform their genetic function, it is clearly necessary that one strand of a DNA double-helix be able to direct the synthesis of the complementary strand. However, it is equally important that the two strands should come apart at the right time. Otherwise, it would be impossible for further DNA replication to occur, or for DNA to direct the synthesis of messenger RNA. It follows that the two strands of a double-helix must be held together by forces that are strong enough to guarantee some stability, but not so strong as to prevent separation at an appropriate moment.

Ordinary chemical bonds *within* molecules are very strong. A double-helix in which the strands were joined by stable chemical bonds could not be separated quickly enough to fulfil a genetic function. On the other hand, the forces *between* molecules are usually so weak and nondirectional that they would be unable to hold a double-helix together in a well-defined structure. DNA is held together by special intermolecular bonds of intermediate strength, known as hydrogen bonds. Each nucleotide in a base-pair forms two or three such bonds, in such a way that the two members of the pair are held in exactly the correct relative configuration.

The structures of the base-pairs are illustrated in Figure 4.4. To define the structures adequately, two or three hydrogen bonds are required in each base-pair. Analogy with a two or three prong electric plug is helpful here. If only one prong is inserted the plug can still rotate in the socket, but once two or three prongs are inserted, the whole structure is held rigidly.

The base-pairs have a remarkable property—they are geometrically equivalent. In Figure 4.5 the thick lines repre-

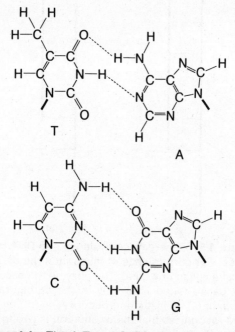

Figure 4.4. The A-T and G-C base pairs. Note the equivalent positions of the bonds joining the bases to the backbone (thick lines).

sent the bonds by which the bases are attached to the back-bone of the double-helix. The hydrogen bonds hold the base-pairs in such a way that these two bonds are the same distance apart and have the same orientation whether we are dealing with an AT, TA, GC, or CG base-pair. It is this property of geometrical equivalence that permits the construction of a regular double-helix from four different components.

We can now see why base-pairing permits accurate replication to occur. The enzyme responsible for joining up the backbone of a new strand of DNA will function only if all new units are presented to it in exactly the same orientation. The preformed DNA strand, acting as a template, presents a

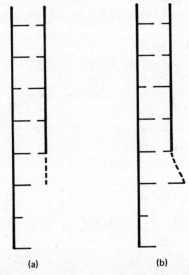

(a)   (b)

Figure 4.5. Base-pair equivalence and DNA replication. For convenience of illustration the double-helix has been unwound. The enzyme will accept a new base (dotted line) only in the configuration shown in (a). If mispairing occurs as in (b), the enzyme recognizes that the orientation is wrong and rejects the incorrect base. (The cell has a second line of defense. If a wrong base slips through, it is usually cut out again.)

nucleotide monomer in the same orientation whether it is T, C, A, or G provided the new unit is base-paired correctly—this is guaranteed by the property of geometrical equivalence. However, if the wrong base attaches itself to the template DNA, it will be presented to the enzyme in a nonequivalent orientation and consequently be rejected. In this way errors of replication are avoided.

It cannot be emphasized too strongly that the ability to form geometrically equivalent base-pairs is a property of the bases themselves, which does not depend on the presence of enzymes or other biological polymers. A forms a specific base-pair with U, and G with C, when the components are dissolved in a simple organic solvent, such as chloroform, for example. As we shall see, it has been possible to make use of the equivalence of the base pairs to imitate some important aspects of nucleic acid replication in totally non-biological systems. It seems likely that a related nonenzymatic replication of nucleic acids was the first "genetic" process to occur on the primitive earth. The evolution of life as we know it was probably dependent on the geometrical equivalence of the Watson-Crick base pairs.

(b) Proteins. A protein consists of one or more polypeptide chains. The sequence of amino acids in a protein is said to define its primary structure. However, one can learn very little about the properties of a protein from the primary structure alone. Usually it is necessary to determine the distribution of the various parts of a protein in space, the secondary and tertiary structures, to understand how a protein functions.

In recent years, X-ray crystallography has been applied successfully to the study of proteins. The structure of a typical protein is illustrated in Figure 4.6. Clearly, only a part of such a large and complicated molecule can be directly involved in the chemical modification of a small substrate. This region is called the active site.

The active site must be confined to a region of space that is not much larger than the part of the substrate that is to be acted on. However, the amino acids that make up the active site are usually widely separated in the primary struc-

69

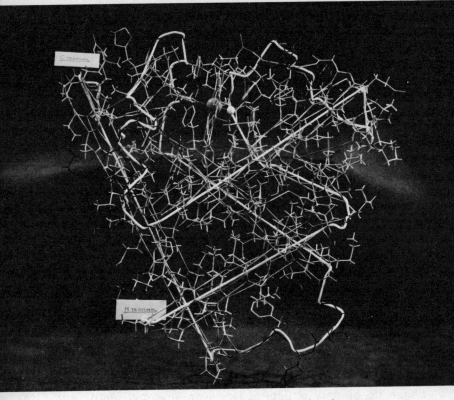

Figure 4.6. (a) Model of Myoglobin showing every atom. The course of the backbone is marked with a white cord. (Reproduced with permission from John Kendrew.)

ture of the protein. The polypeptide backbone is intricately folded to form a structure which places the amino acids of the active site in exactly the correct positions.

Enzymes are not only efficient but also highly selective; they act upon their substrates but often have no effect at all on closely related molecules. The selectivity of an enzyme is controlled by its specificity site. The group of amino acids that make up this site is usually distinct from those in the active site. They are positioned in space in such a way as to direct the appropriate part of the substrate to the active site

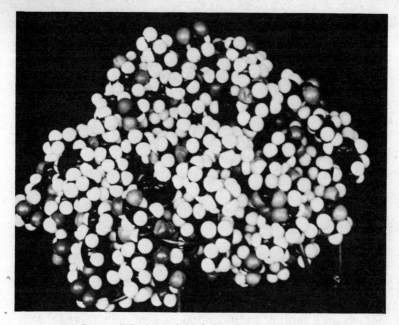

(b) Space-filling model of the same protein. This incredibly complex structure is repeated exactly in every myoglobin molecule. (Reproduced with permission from H. C. Watson.)

but to prevent the approach of molecules other than the substrate.

The extraordinary efficiency with which enzymes catalyze chemical reactions is still not understood completely, but we do know that structural factors are very important. The molecules synthesized by organic chemists are usually quite small, and their functional groups tend to point outwards from a central region. On the other hand, proteins are much larger than their substrates and are often folded so that the amino acids of the active site point inwards from a cavity around the substrate. This permits enzymes to get a very much firmer grip on their substrates than can be achieved by small-molecule catalysts.

The amino acids in the active site of an enzyme are essential for its function; any change in these amino acids usually results in the inactivation of the enzyme. Changes in other parts of an enzyme will often modify its catalytic properties, but will rarely cause complete inactivation. The evolution of enzymes in modern organisms is thought, in most cases, to be associated with changes in the specificity site or in other regions distinct from the active site.

Changes in the specificity site may either increase or decrease the range of substrates on which an enzyme can act. Microorganisms can adapt to use a new sugar as a source of energy, for example, by changing the specificity site of a pre-existing enzyme in such a way as to accept the new sugar as a substrate. On the other hand, they can become resistant to certain drugs by modifying their enzymes in such a way that they no longer act upon the drug. Changes of this type have been important in biochemical evolution.

Changes in the backbone of a protein inactivate an enzyme completely only if they drastically alter the active site. Changes on the outside of a protein, far from the active site, are usually without effect on enzyme activity or affect the activity only slightly. In the course of evolution, it is probably this type of change that has been responsible for those small modifications of protein structure that adapt an enzyme to work under slightly changed conditions. The adaptation of an enzyme to work at somewhat higher or lower temperatures, for example, is believed to occur by the accumulation of changes in the primary sequence outside and possibly far away from the active site.

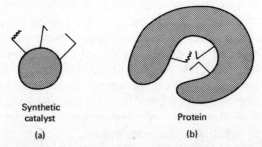

Synthetic
catalyst

(a)

Protein

(b)

Figure 4.7. (a) A synthetic catalyst with functional groups pointing out. (b) A much larger protein catalyst with functional groups that can encircle a substrate.

Chemical Free Energy

Most familiar chemical reactions are spontaneous. They are also irreversible in the sense that they do not proceed spontaneously in the opposite direction. A piece of iron rusts spontaneously, but rust never spontaneously reverts to metallic iron; coal burns to carbon dioxide and water, but a mixture of carbon dioxide and water has no spontaneous tendency to form organic material again. Quantitative studies of systems of this kind have contributed to the development of a branch of chemistry called chemical thermodynamics. The notion of equilibrium plays a central part in this subject.

A *closed* chemical system is said to have reached equilibrium if it does not tend to undergo further chemical change. Iron exposed to damp air, for example, does not reach equilibrium until it is completely converted to rust. Our definition implies that any *closed* chemical system would reach equilibrium if left long enough. It continues to react so long as it has any tendency to react; when it has no further tendency to react it is, by definition, in equilibrium.

This definition of equilibrium contains no reference to time. In some cases the approach to equilibrium is so slow that it cannot be studied in the laboratory. Nonetheless, it is always possible, in principle, to calculate the properties that a system would have in equilibrium, if it had time to get there. Alternatively, special methods can be used to accelerate the approach to equilibrium. A mixture of hydrogen and oxygen reacts very slowly at room temperature, but it equilibrates explosively if exposed to a small spark. Enzymes and other catalysts accelerate the approach to equilibrium, but in a less dramatic way.

The nature of equilibrium in a chemical system is made clearer by considering the behavior of more familiar mechanical systems. Water flows spontaneously downhill, but never flows spontaneously uphill. The state in which no further motion occurs because the water has reached the lowest level attainable corresponds to the equilibrium state in a chemical system. An isolated system containing parts that move relative to each other sooner or later comes to rest under the influence of friction; however, systems never speed up under the action of friction (unless external forces are applied). In this case the "equilibrium" state is the one in which there is no tendency for the relative velocities to change — in a system subject to frictional forces it is the state in which the parts are at rest relative to each other.

These ideas lead us naturally to the idea of free energy. Man has learned to exploit many of the spontaneous processes that occur in his environment to help him to do work. Water flowing downhill is used to turn paddlewheels; the motion of the air is harnessed and made to push sail boats; fuel is burned and the heat that is evolved is used to drive engines. We use the term free energy to describe the part of the energy released in spontaneous processes that can, in principle, be used to do work.

The free energy is the upper limit of the amount of work that can be derived from a given spontaneous process. It is possible to build inefficient machines that waste most of the available energy, for example, by converting it to frictional heat, but no machine can be constructed that does more work than that which corresponds to the available free energy. Real machines are always somewhat inefficient owing to friction, etc., but it is usually possible to calculate how much work could be done by a hypothetical, perfectly efficient, frictionless machine.

When we do work on a mechanical system we make use of the free energy released in one process as it takes place spontaneously to drive another process away from mechanical equilibrium. In typical cases we might use water power to raise a weight or electrical power to accelerate the air in a wind tunnel. In all such cases we couple together two systems so that the free energy that is released as one moves spontaneously towards equilibrium drives the other one away from equilibrium.

Now let us return to chemical systems and compare them with the mechanical systems discussed above. Work can be derived from chemical systems if they are capable of reacting spontaneously, that is if they are not already in equilibrium. The chemical free energy corresponding to any given reaction is the maximum amount of work that can be obtained from the reaction as it proceeds spontaneously to

The Nature of the Problem

equilibrium. Just as with mechanical systems, the maximum work that can be realized from a chemical reaction can usually be calculated, even though it can rarely if ever be realized.

The free energy that is available from spontaneous chemical processes can be used in a variety of ways. Hydrocarbons can be burned and the heat that is released at high temperatures can be used to drive a steam engine, for example. In biological systems chemical reactions are almost always coupled together directly so that the free energy released in one reaction as it moves towards equilibrium drives another reaction away from equilibrium. We shall see that in cells part of the free energy available from the oxidation of glucose is used to synthesize ATP. The free energy that could be derived from the hydrolysis of ATP is then utilized in turn to bring about the synthesis of proteins, nucleic acids, and other molecules.

Heat engines are almost always a good deal less than 100% efficient, so that the work done is always less than the free energy that is available. The same is true of biochemical "engines"; only about half of the energy available from the oxidation of glucose, for example, is converted to ATP. Nonetheless, the enzyme-catalyzed energy conversions that occur in cells are still a good deal more efficient than are similar processes in nonliving systems.

Free Energy in Biological Systems

We have seen that all *closed* chemical systems ultimately run downhill to equilibrium. Living systems somehow escape this fate. This once led some authors to believe that living systems are guided by some "vital force" that enables them to evade the laws of thermodynamics. The situation is now realized to be much simpler. Living organisms are not *closed* systems, so the laws of thermodynamics in their standard form do not require that they run downhill to equilibrium. The following analogy may be useful. An airplane with a fixed supply of fuel must ultimately make contact with the earth, but an airplane that is refueled in mid-air can fly forever (leaving aside wear due to friction).

A population of bacteria such as *E. coli* can divide forever, so long as it is constantly "refueled" by glucose or some other nutrient in the environment. The bacterium avoids running downhill to equilibrium as long as it can derive energy from an external supply of organic compounds. There is no "thermodynamic mystery" about the survival and reproduction of simple living organisms as long as they are adequately nourished.

However, if we consider all of the interdependent organisms that make up the biosphere, we do have to explain how they get their energy. Living organisms, like other chemical systems, are constantly dissipating free energy. If this free energy could not be replaced, all life would inevitably come to an end. There is only one important process that restores free energy to the biosphere—photosynthesis. Insofar as energy is concerned, the biosphere is living on foreign aid; photosynthesis is the only exchange mechanism that permits light emitted by the sun to contribute to the maintenance and evolution of life on earth.

A single substance, ATP, is involved in almost all energy transactions within the cell—pursuing the analogy with economics, we may regard ATP as the universal currency of biological energy exchange. Different organisms derive their

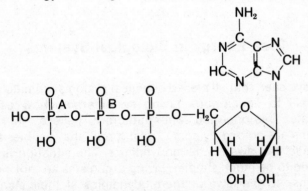

Figure 4.8. The structure of ATP. The bonds marked A and B are those which are split when ATP is used to provide energy for biochemical reactions.

The Nature of the Problem

energy from different sources, but all of them convert it to ATP as a preliminary to utilizing it for growth and function.

Among the many biochemical reactions that generate ATP, those that proceed in the absence of oxygen are most interesting to us, since they could have occurred on the primitive earth. These reactions are often referred to as fermentations. In fermentations, the atoms that make up organic molecules, such as glucose, are rearranged in a complicated way to form more stable compounds; energy that would be released as heat during the rearrangement, if it occurred spontaneously in a nonbiological system, is used to form ATP by living organisms.

Higher animals use a quite different mechanism, oxidative phosphorylation, to obtain most of the energy they need. The oxidation of glucose to carbon dioxide by oxygen from the air is coupled to the synthesis of ATP. Much more energy is released by the oxidation of glucose than by fermentation; so, much more ATP is formed for each molecule of glucose used up. However, oxidative phosphorylation must be a relatively recent biological innovation, since, as we shall see later, there was no oxygen in the primitive atmosphere.

The free energy content of ATP is stored in the bonds that join together the three phosphorus atoms. These bonds are, therefore, called high-energy phosphate bonds. When ATP is left in contact with water, the high-energy bonds are broken and the free energy stored in them is wasted. In the presence of suitable enzymes, however, these bonds are not hydrolyzed directly; instead, ATP undergoes a series of coupled reactions that lead to changes useful to the cell. The final products obtained from ATP are the same as those obtained by direct hydrolysis, but part of the energy which would be released as heat in direct hydrolysis is made to do useful work in biological systems.

Free Energy and the Synthesis of Proteins and Nucleic Acids

We have seen that the survival of living organisms depends on their ability to use external sources of energy to produce ATP. The steps by which energy is used to synthesize ATP are poorly understood. We are better informed about the methods that are used to make ATP do useful work, for example, in bringing about the synthesis of biological polymers. The following treatment of this complicated subject is much oversimplified.

In Chapter 9 we shall see that biological polymers are never in equilibrium with water; an aqueous solution of a protein, for example, tends to decompose into amino acids. It follows that protein synthesis, because it converts amino acids to proteins, is an uphill process and must be driven by the decomposition of ATP. How are two such different reactions as the downhill hydrolysis of ATP and the uphill synthesis of proteins coupled together?

The general principle involved is fairly straightforward. If ATP is to influence the behavior of amino acids, it must first react with them. Furthermore, during the reaction as much as possible of the free energy originally stored in the ATP must be conserved, for otherwise no energy would be left to drive the synthesis of proteins uphill. Stated in another way, ATP must first react with the amino acids to form high-energy intermediates that preserve most of the free energy released by the decomposition of the ATP.

The necessary reactions are catalysed by a set of enzymes (activating enzymes) that are universally distributed in living organisms.

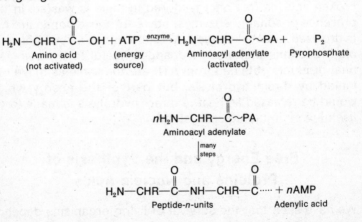

Net Result: nATP + nAmino Acid $\longrightarrow$ nPeptide + nAMP + nPyrophosphate

Once the activated intermediates are formed they could, in principle, react with one another directly to form the peptide bonds of proteins. The energy released by the breakdown of the intermediates is sufficient to drive the synthesis of proteins uphill. In practice, the mechanism of protein synthesis is more complex (see below), but the general principles involved are similar.

It is thought that most, if not all, biological condensation reactions proceed by somewhat similar mechanisms. We know less about prebiotic condensation reactions. It is unlikely that ATP was the first prebiotic source of energy, but somewhat similar molecules may have been available on the primitive earth. In any case, it is almost certain that many prebiotic condensation reactions involved activated intermediates of one kind or another (Chapter 9).

Photosynthesis

An important consequence of photosynthesis is the formation of organic materials at the expense of sunlight and carbon dioxide in the atmosphere. In plant photosynthesis, water is the source of the hydrogen needed to reduce carbon dioxide to sugars; oxygen is released into the atmosphere.

$$6CO_2 + 6H_2O \longrightarrow (CH_2O)_6 + 6O_2$$
$$\text{Glucose}$$

Carbon dioxide can also be reduced by microorganisms at the expense of other substances such as hydrogen sulfide. Since these alternative reducing agents may have been abundant on the primitive earth, they may have been significant for the evolution of photosynthesis.

Plant photosynthesis provides the organic compounds that other organisms use, directly or indirectly, as sources of structural material and energy. The higher animals, for example, recover the energy stored by plants when they oxidize the products of photosynthesis back to carbon dioxide, using atmospheric oxygen. The oxygen itself has, of course, been generated by photosynthesis.

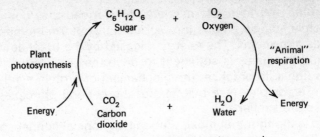

Figure 4.9. The biological carbon cycle.

Protein Synthesis

Protein synthesis is a complex process in which amino acids are joined in specific sequences according to instructions encoded in a stretch of DNA. The nucleotides making up the DNA are read in groups of three from a fixed starting point. The amino acid assigned to any group of three nucleotide bases is specified by the genetic code (Figure 3.14).

Each segment of a DNA double-helix consists of two complementary strands. Of these, only one is used to specify the sequence of amino acids in proteins. We shall refer to this strand as the positive strand. We have seen that the DNA does not take part directly in protein synthesis, but is represented by messenger RNA. Clearly, the messenger RNA must have the same sequence as the positive strand of DNA, and hence it must be constructed using the negative strand of DNA as template.

The assembly of proteins takes place in or on a ribosome, a complex structure composed of proteins and nucleic acids. Messenger RNA and "prepared" amino acids associate with the ribosome—unchanged messenger RNA and proteins are released.

A remarkable feature of protein synthesis is the "preparation" that precedes that presentation of amino acids to the ribosome. Amino acids do not enter the ribosome in the free form, nor as any simple, activated derivatives. Instead, each amino acid is first attached to its own special type of RNA molecule. These RNA molecules are called tRNA's (short for transfer RNA's); the tRNA to which glycine becomes at-

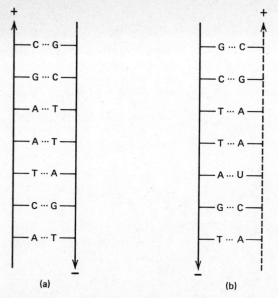

Figure 4.10. Solid lines represent DNA strands; dotted line represents messenger RNA: (a) resting double-strand of DNA; (b) negative DNA strand directs synthesis of positive RNA strand.

tached is written as tRNA$_{gly}$, and similarly with the other amino acids.

The attachment of amino acids to their tRNA's thus constitutes a preliminary sorting of amino acids — glycine is attached to tRNA$_{gly}$, alanine to tRNA$_{ala}$, and so on. This sorting is achieved with the help of twenty activating enzymes, each of which recognizes both a tRNA and the corresponding amino acid. The process of attaching an amino acid to its tRNA is referred to as loading.

The tRNA's have another remarkable property: they form complicated three-dimensional structures often referred to as cloverleaf structures, in which special sequences of three residues are exposed in just the right way to form base-pairs with three nucleotides in messenger RNA. The choice of these three exposed bases on the tRNA's is a key aspect of protein synthesis. Let us see how this works out in a simple case.

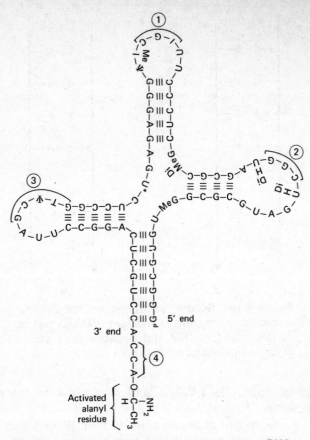

Figure 4.11. (a) The structure of alanine tRNA drawn in the conventional cloverleaf form. (Reproduced with permission from Holley et al., *Science*, **147**, 1462 (1965).)

We know that glycine is incorporated into a protein sequence whenever it is signalled by the sequence GGG in a messenger RNA. This translation is made possible by a tRNA$_{gly}$, in which the exposed sequence is CCC. The tRNA$_{gly}$ is first loaded by its activating enzyme; next the exposed CCC sequence of the tRNA$_{gly}$ associates tightly with the GGG sequence in the messenger RNA by Watson-Crick base-pairing. This guarantees that a glycine is placed in the right

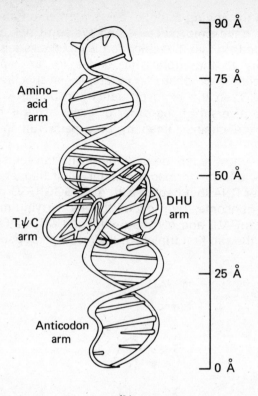

Amino–
acid
arm

DHU
arm

TψC
arm

Anticodon
arm

90 Å

75 Å

50 Å

25 Å

0 Å

(b)

(b) One of the proposed fully folded structures of tRNA. Note how arms two and three are folded around the waist of the molecule. (Reproduced with permission from A. L. Lehninger, in *Biochemistry,* Worth Publishers, Inc., New York, 1970.)

position to be incorporated into the protein opposite the GGG sequence in messenger RNA.

This example illustrates the general principle. Each amino acid can be loaded onto one or more tRNA's. These tRNA's have exposed sequences of bases that pair with exactly those triplets of bases in the messenger that code for the amino acid in question. The exposed triplet of bases on a tRNA is called an anticodon, because it base-pairs specifically with the coding triplet of bases (codon) on the messenger RNA. In the simplest cases only Watson-Crick base-pairs are involved, so that the anticodon is just the sequence complementary to the triplet that codes for the appropriate amino acid. Tryptophan, for example, is specified by the triplet UGG, so the exposed anticodon of tRNA$_{try}$ is CCA. (Note the reversal of the order of bases. This is necessary because the chains joined by base-pairs run in opposite directions.)

To see how a peptide beginning with the sequence met-gly-try... could be formed, consider what happens when a messenger RNA that begins with AUG GGG UGG associates with the ribosome. Two loaded tRNA's carrying methionine (anticodon CAU) and glycine (anticodon CCC) attach themselves to the two first triplets of the message. A peptide bond

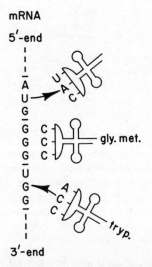

Figure 4.12. A diagrammatic representation of the formation of a polypeptide which begins with the sequence met-gly-tryp. . . . The first tRNA which carried the methionine has fulfilled its function and is shown falling off the messenger. The third amino acid (tryp) is just taking its place on the messenger. In the next stage met-gly will be transferred to the loaded tryp tRNA.

is then formed leaving a tRNA$_{gly}$, loaded with met-gly, on the ribosome; at the same time the unloaded tRNA$_{met}$ falls off. Next a loaded tRNA$_{try}$ (anticodon CCA) associates with the ribosome and attaches itself to the UGG sequence; this time the peptide met-gly is transferred, leaving the peptide met-gly-tryp attached to tRNA$_{try}$ and freeing the tRNA$_{gly}$. A long protein chain can be built up in a sequence of simple steps of this kind.

The tRNA's are often referred to as adaptors, because they make it possible for messenger RNA to order the amino acids in a protein, without ever coming into direct contact with them. This is an extremely important point. Messenger RNA's are the kind of molecule that can associate specifically with other RNA molecules by base-pairing, but they cannot discriminate directly between amino acids. The use of tRNA's allows this difficult discrimination to be carried out by a protein. The messenger RNA has only to order the correct tRNA's, for everything else is done by the activating enzymes.

At an early stage in the evolution of life, before the development of protein synthesis, no activating enzymes with *highly specific* catalytic activity could have existed. How could polynucleotides have directed the synthesis of polypeptides, before the evolution of the activating enzymes? This is one of the great puzzles of evolutionary biology. We shall return to it in Chapter 10.

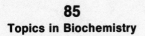

The Biochemical Record

5

Paleontologists, by studying the fossil record, have traced back the history of life some three billion years. Unfortunately, no corresponding biochemical record is available. While fossils often survive for a very long time, most organic chemicals in the fossils are destroyed quickly by heat or moisture. Thus, although the study of the organic substances associated with fossils is progressing, it is unlikely to tell us much about the very earliest phases in the development of life.

We are forced to conclude, therefore, that the only evidence that the biological sciences are likely to provide about biochemical evolution must come from the examination of species that are still alive today. The task of reconstructing the history of the origins of life from the biochemistry of living species can be compared with that of establishing the history of technology by examining machines constructed since 1970. How, in the latter case, could one tell whether the use of the wheel and the lever were contemporary innovations, or whether one was introduced into a technology that already made use of the other? No doubt, there are questions about the origins of life which will turn out to be of this kind; that is, questions that are unanswerable unless we can find the historical evidence. Fortunately, some of the questions that are most interesting to us can be answered in other ways.

Valuable information about the early stages in biochemical evolution can be obtained by comparing the organic constituents of different modern organisms. If we find the same organic compound in members of two living species, we have to decide between two alternatives. The common constituent may occur in the two species because the ability to make it has been inherited from a common ancestor, or the steps needed to synthesize the substance may have developed independently in the two species in response to some common need. We refer to these two processes as divergent and convergent evolution, respectively. While it is sometimes difficult to decide between these alternatives, in the most important instances the choice is obvious.

If, for example, the synthesis of some complicated organic chemical evolved independently in two species, it is unlikely that the two methods used to make it would be identical. Convergent evolution, therefore, can explain the similarities between related chemical pathways in different species only if the degree of similarity between the pathways can reasonably be attributed to chance. If not, we must assume that divergent evolution has occurred. Compare the problem of deciding between convergent and divergent evolution with that of discovering whether or not the work of two inventors is independent. Clearly, if the inventions are very similar and a whole paragraph appears in identical form in their patent applications, we are justified in supposing that one is a plagiarist (or both, if they have a common "ancestor"). The identity of passages in the patents cannot be attributed to chance and, therefore, proves that the inventors did not "converge" on the same solution to their problem.

Very complicated features of biochemistry which are identical in sufficient detail in two species must, therefore, have been inherited from a common ancestor. It follows that if certain complex features of biochemistry are common to all forms of life, then there must have been a common ancestor from which all living things are descended.

We have already seen that the genetic apparatus is a very complex but universal feature of all living cells. The components of DNA and RNA, the 20 amino acids, and the genetic code are the same in all organisms. It is inconceivable that two systems that evolved independently could resemble one another so closely, so we may safely conclude that there was a common ancestor of all living things.* We shall refer to this organism as our last common ancestor.

Our conclusion that there must have been a unique, complicated ancestor of all living things poses a new problem. Clearly our last common ancestor must have been the product of a long evolutionary process. Why do we not see evidence of other forms of life that evolved in parallel with it? The nature of the problem and an outline of an answer are indicated in Figure 5.1.

In the course of time, any "species" gives rise through mutation and natural selection to new and competing "species." Some of these ultimately become extinct, but the successful ones go on in their turn to produce even more varied forms of life. The natural consequence of evolution is the production of different "species," each adapted to its own special environment. What happened to the organisms which were competing with our last common ancestor?

To understand the complete dominance of the descendants of any organism, we must suppose that it acquired so considerable an advantage over all competitors that it was able to eliminate them. It is unlikely that we shall ever know the nature of the "invention" that made this possible. It could, for example, have been an extracellular enzyme able to attack all competitors, or a more efficient method of producing ATP. Possibly it involved the pooling of "inven-

* Strictly, we can deduce only that all modern organisms are derived from a small inbreeding ancestral population.

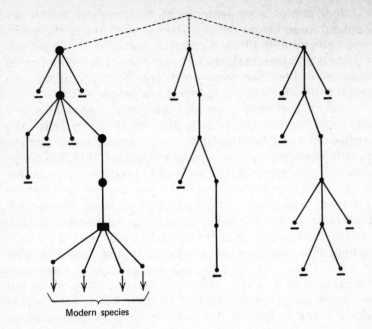

Modern species

● = Common ancestor
■ = Last common ancestor
— Denotes extinction of "species"

Figure 5.1. Our last common ancestor. It is assumed arbitrarily that there were three independent "families" of organisms. Note that several organisms coexisted with out last common ancestor but were finally eliminated by species that diversified from our last common ancestor. (The dotted lines indicate that the three families under consideration could have had a common origin even further back in the past.)

tions" between two organisms in some form of genetic exchange, rather than a change within a single organism. In any case the biochemistry of the organism in which the crucial advance was made ultimately became the universal form of biochemistry.

If we accept these arguments, we must conclude that

The Nature of the Problem

comparative biochemistry alone can tell us little about the origins of life. In the absence of additional evidence or some more subtle theoretical arguments, there is no way of learning about any organism more primitive than our last common ancestor. All traces of other forms of life that must have coexisted with it and differed significantly from it have disappeared from the biochemical record.

We have at last come face-to-face with the irreducible problem of the origins of life. How did an organism that already incorporated all of the universal features of contemporary biochemistry develop from the inorganic constituents on the surface of the primitive earth?

If no trace remains of primitive forms of life dating back to the period before the development of our last common ancestor, what other evidence is available to throw light on the earlier phases of biochemical evolution? The only way in which we can derive useful information would seem to be by studying the organization of biochemical reactions in modern cells. What we need is a speculative reconstruction of the sequence of steps involved in the evolution of biochemistry and in particular in the evolution of the genetic apparatus.

This rather difficult idea is best illustrated by a simple example. In all present-day organisms the genetic apparatus is made up of proteins and nucleic acids. What reason is there to believe that the first genetic apparatus contained nucleic acids? In Chapters 3 and 10, it is emphasized that proteins cannot replicate and, hence, that a genetic system cannot be constructed from proteins alone. Thus, we shall see that, even in the absence of direct fossil evidence, we can be almost certain that the first organisms contained nucleic acids or closely related polymers.

We should be clear, however, that the study of the origins of life, because it is a historical study, is attended by difficulties that do not occur in most other branches of science. We are searching for a historical reconstruction that must be imaginative, but not too imaginative. What constraints are needed to guarantee that speculation is kept within reasonable bounds?

The answer to this question is quite straightforward.

Each step in any proposed history of the evolution of life must be compatible with all relevant scientific knowledge. Chemistry, geology, and astronomy all provide information about the primitive earth and the processes that could have occurred on it. Finding a theory that is plausible in the light of all the relevant information will certainly be very difficult. It seems unlikely that we shall ever find two radically different theories, both of which are satisfactory.

Any complete theory of the origins of life must specify a series of reactions that could have occurred under prebiotic conditions and it must be adequate to explain the evolution of our last common ancestor. A theory gains in status with each new demonstration that a proposed step can be simulated plausibly in the laboratory. Since our last common ancestor was composed of organic molecules similar to those occurring in modern organisms, it is first necessary to show how these substances were formed on the primitive earth. Then we must explain how a primitive organism evolved from a mixture of suitable organic molecules.

Unfortunately, the conditions that existed on the primitive earth are very different from those that are usually used by organic chemists. Although many of the constituents of cells can be synthesized in the laboratory, a few of the most important of them cannot yet be made under prebiotic conditions. It is even harder to join the simple organic molecules together to form polymers similar to proteins and nucleic acids. Consequently, a great deal of work will have to be done before we can propose a single complete theory of the origins of life and show by experiment that each step could have occurred on the primitive earth.

Our Program

It is important at this point to distinguish two aspects of the problem of the origins of life. In the next section of this book we shall be concerned with the origins of life on earth. This is a very difficult subject, but the difficulties are not of a philosophical nature. We have to explain the evolution of cells from inorganic matter; we do not have to define "life" in an abstract way, neither do we have to decide at what point the word "living" could first be applied to the evolving system of organic molecules on the primitive earth.

In Chapters 13–16 we shall speculate more generally about life in the Universe. Then we shall have to consider questions that have a philosophical or at least a semantic aspect. What is life? Need it be based on the chemistry of carbon? Need it be based on "chemistry" at all? It is fortunate that we do not have to answer these more elusive questions before we can get started on the topic that interests us most.

For the moment we have only to explain the evolution of our last common ancestor. Clearly, like a modern cell, our last common ancestor was a complicated chemical factory enclosed in a lipid membrane. All important chemical reactions were catalyzed by proteins made of the twenty standard amino acids. Protein synthesis was directed by nucleic acids indistinguishable from those in modern cells. Even the complex system of enzymes and nucleic acids that is responsible for modern protein synthesis had achieved its final form.

The synthetic ability of our last common ancestor was equally impressive. Mechanisms had already evolved that permitted the synthesis of the standard amino acids, sugars, and nucleotide bases from very simple starting materials. In addition, the pathways leading to most other important biochemicals were available. ATP could be generated by the decomposition of sugars and used to support many kinds of synthetic activity. Recent evidence suggests that the sequences of amino acids in many of the enzymes present in our last common ancestor were very similar to those of the corresponding enzymes of modern organisms.

It will be convenient to divide our discussion of the origins of life into three sections, corresponding to three major phases in the evolution of our last common ancestor. First we shall describe what is known about the primitive earth and about the prebiotic synthesis of organic material on its surface. We shall be most interested in the components of the genetic apparatus, the amino acids, sugars, and nucleotide bases. This aspect of prebiotic chemistry is discussed in

Chapters 6–8. Next, in Chapter 9, we go on to consider how the small molecules, once they had accumulated in sufficient quantities, were joined together to form random nonbiological polymers. Finally, in Chapters 10 and 11, we come to the most difficult but also the most interesting part of the problem: how did a highly organized cell evolve from a mixture of random organic polymers?

Appendix to Chapter 5 — Panspermia

The account of the origins of life which we have outlined is the conventional one. However, a quite different theory was popular during parts of the nineteenth century. In Chapter 1 we discussed briefly a theory called Panspermia, according to which life did not evolve from inorganic matter on earth, but reached us fully developed in the form of a bacterial spore that had escaped from a distant planet. The theory, as originally proposed, had mystic overtones, but the idea that life evolved elsewhere in the universe and then infected the earth is not in itself unreasonable. I do not think that any theory of this general kind is likely to be correct, but since such theories have often been dismissed too dogmatically, they are worth discussing here.

It seems certain that no spores could survive the journey from another solar system to the earth. It is easy to calculate the amount of radiation that a spore would receive during the journey, and this is many orders of magnitude greater than that needed to kill a terrestrial spore. More significantly, the amount of radiation received would seriously disrupt any organized material made up of carbon, hydrogen, nitrogen, and oxygen. Thus, the theory of Panspermia in its strictest form cannot be correct. However, the related theory, that the first cell arrived on the earth within a meteorite, is harder to disprove.

The frequency with which meteorites reach the earth from distant parts of the galaxy could be calculated if we knew how often chunks of matter escape from planetary systems. Unfortunately this information is not available. Estimates that have been made recently suggest that escape is extremely improbable and hence that very few meteorites make the journey. The difficulty is that one might have been enough, and no one can prove that in the first billion years of the earth's history no single object from outer space penetrated our atmosphere.

If a living cell escaped from another planet within a meteorite, there is no reason to doubt that it could have survived the journey to the earth. It would have been protected from radiation by the solid material of the meteorite, while the low temperature of outer space would have prevented any spontaneous chemical deterioration. Chapter 16 presents arguments which suggest that there may be a class of planets with evolutionary histories similar to that of the earth. An organism from such a planet might well have found the earth a hospitable place.

The assumption that life was brought to the earth from another planet would deserve more attention if it made it easier to understand how life evolved, or if it explained some surprising facts about life on earth. Is this the case? Clearly if we believe that life began on another planet, we are justified in considering a wider range of prebiotic environments than could have existed on the earth. It could be imagined, for example, that life began in an ocean of liquid ammonia, since it is quite possible that there are planets which have oceans of this kind. It turns out that this additional freedom in choosing a prebiotic environment would not be very helpful. The conditions that we customarily assume to have existed on the primitive earth are as favorable for the origins of life as any that can reasonably be postulated for another planet.

The situation is rather different when we come to consider the evolution of biological order. Many molecular biologists are unconvinced by the arguments that we have given to explain why biochemistry is so uniform. They find the universality of the genetic code particularly surprising, since it is hard to see why competing organisms with slightly different codes could not coexist. If we could show that life evolved on another planet, this difficulty would be removed.

Suppose that life on some other planet had evolved to produce organisms much more diverse than those on earth. On such a planet the genetic code might well have been different in different species. However, if a meteorite carried a single organism to the as yet lifeless earth, that organism would have taken over completely before any competitors could appear. Thus, all terrestrial organisms would have inherited the particular code used by their unique extraterrestrial ancestor. In biological terms all terrestrial life would consist of a single family derived from a more diverse population on another planet.

I think it more likely that a unique code evolved on the earth than that life reached us from another planet. However, the latter point of view is an interesting one and should be entertained by students of the origins of life, at least on sleepless nights. Could life have been planted on the earth, deliberately, by the members of a technological society on another planet?

PART
TWO

STEPS
TOWARD A
SOLUTION

History of the Earth, Atmosphere, and Oceans

6

Introduction

A relatively small number of simple organic molecules are involved in the fundamental biochemical processes of protein synthesis and nucleic acid replication. The same twenty amino acids occur in the proteins of all organisms. The nucleic acids isolated from cells always contain the same sugars, ribose and deoxyribose, and the same set of nucleotide bases. If the earliest organisms were recognizably related to contemporary organisms, they too must have made use of some of these compounds. The demonstration that many biologically important compounds can be formed from inorganic material under conditions similar to those that must have prevailed on the primitive earth is perhaps the most important advance we have made so far in our understanding of the origins of life. Before we can begin to describe this work we must outline what is known about the

99

early history of the earth, for it was in the atmosphere, oceans, and lakes of the primitive earth that the chemical constituents of the first living organisms were made.

Origin of the Solar System

It was once widely believed that the solar system was formed when another star passed very close to the sun; the planets were thought to have been torn from the sun by the very large gravitational field of the nearby star. This theory is no longer tenable. Calculations show that it is necessary to assume an extremely close and therefore extremely improbable near-collision to account for the formation of objects as large as the planets. Furthermore, even if objects that large had been formed, they would have had so much internal energy that they would have disintegrated.

All modern theories of the formation of the solar system are based on the Kant-Laplace hypothesis, according to which the sun and planets are derived from a vast cloud of dust and gases. Many such dust clouds have been observed in our own galaxy. A dust cloud, as it rotates, first flattens into a disc. If the gravitational field is large enough, material then begins to accumulate at the center of the dust cloud. In time, as more and more material condenses, a central star is formed.

However, the accumulation of material at the center of the dust cloud does not continue indefinitely. The material in the dust cloud is rotating around its center; consequently, as the cloud collapses the rate of rotation must increase. If all the material formed a single star, the star would spin so quickly that its outer layers would be thrown off again. Instead most of the rotational motion in the dust cloud is transferred to a relatively small amount of matter which subsequently moves in orbit around the central star.

It is believed that the material in the solar system which escaped the initial condensation and remained in regions far from the sun took part in a series of secondary condensations. Solid objects were formed and gradually swept up all the dust particles in concentric rings around the sun. These

Figure 6.1. An imaginative illustration of the later stages in the condensation of the solar system according to the Laplace hypothesis. (Reproduced with permission from *Stars, Planets and Life* by Robert Jastrow, William Heinemann, Ltd., London, 1967.)

objects became the planets. The Kant-Laplace hypothesis explains in a simple way why the planets rotate about the sun in a single plane and why the plane of rotation of the sun coincides with the plane of rotation of the planets around the sun. Once the dust cloud had flattened into a disc, all further rotational motion had to take place within the plane of the disc.

The composition of the planets must have been determined by the density, composition, and temperature of the dust cloud in the regions where they condensed. There is little doubt that the original dust cloud contained a large excess of hydrogen and helium, since these elements are so much more abundant in the universe than are any of the heavier elements. The major planets, Jupiter and Saturn, do indeed contain a great excess of hydrogen and helium, but the composition of the earth is very different. The earth contains large amounts of heavier elements, such as oxygen and iron, and very little hydrogen and helium. Why did hydrogen and helium escape from the earth?

A gas can be retained in the atmosphere of a planet only if the gravitational field of the planet is sufficient to prevent molecules of the gas from escaping into space. Three important factors determine which molecules can escape and which are trapped in the atmosphere. The heavier the planet, the larger the gravitational field. Hence, it is harder for atoms or molecules to escape from larger planets. The colder the atmosphere, the slower the motion of individual atoms or molecules. Hence, it is harder for atoms or molecules to escape from a cold atmosphere. Finally, the heavier the atoms or molecules, the more strongly they are held back by the gravitational field. Hence, light atoms and molecules escape more easily than do heavy ones.

A complicated mathematical theory has been developed to make these ideas more quantitative. It is found that the major planets, Jupiter and Saturn, are large and cold enough to retain all substances in their atmospheres, including hydrogen and helium. On the other hand, Mars and the earth, because they are much smaller and hotter, lose the lightest atoms, hydrogen and helium, although they can retain molecules such as nitrogen, oxygen, and water. The moon, which is much smaller again, cannot retain any atmosphere.

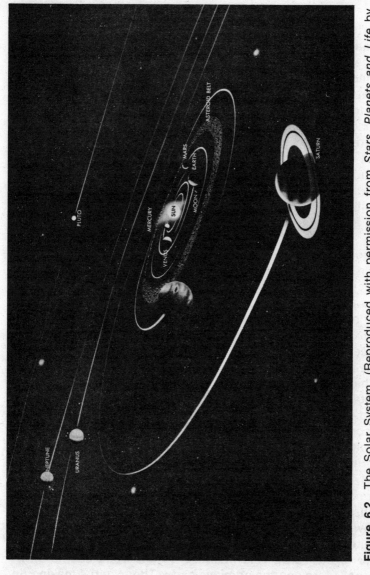

Figure 6.2 The Solar System. (Reproduced with permission from *Stars, Planets and Life* by Robert Jastrow, William Heinemann, Ltd., London, 1967.)

History of the Earth

Two methods have been used to estimate the age of the earth. One of them depends on the assumption that the planets and the meteorites were formed at about the same time. The ages of a number of meteorites have been determined by radioisotope methods. There is general agreement that the meteorites were formed about 4.5 billion (4.5×10^9) years ago. The second estimate of the age of the earth depends on measurements of lead isotope ratios in various terrestrial rocks. The detailed arguments are complicated, but the conclusion is quite straightforward. The lead isotope ratios also point to an age of about 4.5 billion years for the earth. In view of the close agreement between these two quite independent estimates, an age of 4.5 billion years can be accepted with a good deal of confidence.

The temperature of the surface of the primitive earth is also of great concern to us, but in this case we are less certain about the facts. The widespread belief that the earth was once completely molten is based on the theory that the planets were torn from the sun in a near-collision with another star. Since the collision theory has been abandoned, there is no longer a valid reason for believing that the surface of the earth was ever very hot. It has been claimed that the presence in the earth's crust of a number of substances which would have been driven out by temperatures above 300°C proves that the surface of the earth was never very hot. This argument is not universally accepted, and some geophysicists believe that the surface of the earth was melted by the heat released within the earth soon after its formation.

While this controversy remains to be resolved, there are good reasons for believing that volcanic activity on the earth must once have been a good deal more extensive than it is now. Volcanic action at present is sustained entirely by heat generated by the decay of radioactive substances in the interior of the earth. This was also an important source of internal energy on the primitive earth. The material making up the primitive earth certainly contained larger amounts of the radioactive elements than are present now. It is known, for example, that most of the radioactive potassium isotope, ^{40}K which was present on the primitive earth has already decayed; so, energy production from ^{40}K was many times greater on the primitive earth than it is now. The same considerations apply, more or less strongly, to the production of energy by each of the other radioactive elements.

Since energy released in radioactive decay is the only energy available to maintain volcanic activity, and since much more of this energy was released on the primitive earth than is released now, it may be safely assumed that volcanic activity was once more extensive than it is now. However, it is hard to be sure just how large an effect this was. It is likely that, despite the much higher level of volcanic activity, the average temperature of the earth's surface soon settled down within the range that prevails today. In all subsequent discussion of the origins of life, it will be assumed that the temperatures on the primitive earth were similar to those on the contemporary earth.

We have seen that the absence of hydrogen and helium in the earth's atmosphere is readily explained by the small size and relatively high temperature of our planet. When the abundances of certain other elements are examined, a paradoxical situation is revealed. Both neon and argon have been lost almost completely from the earth's atmosphere, but water, nitrogen and, oxygen have been retained. Theory shows clearly that since water, nitrogen, and oxygen mole-

Table 6.1 Atomic or Molecular Weights of Constituents of a Reducing Atmosphere

Atom or Molecule	Formula	Mass
Hydrogen atom	H	1
Hydrogen molecule	H_2	2
Helium	He	4
Methane	CH_4	16
Ammonia	NH_3	17
Water	H_2O	18
Neon	Ne	20
Argon	A	40

cules are lighter than argon atoms, they should have escaped more easily than argon.

There is only one reasonable explanation of this unexpected result. At a time when the earth had only partially condensed, neon and argon were free in the atmosphere and, hence, could escape from the weak gravitational field, while carbon, nitrogen, and water were present in nonvolatile materials trapped in the dust particles. This explanation is a very reasonable one. The outstanding feature of the chemistry of neon and argon is that these gases do not form any chemical compounds at all. Unlike carbon, nitrogen, and oxygen, they could not have been trapped within the dust particles in a chemically combined form.

We are thus forced to a surprising and important conclusion, namely, that the gases in the atmosphere and the water in the ocean were expelled from the interior of the earth. When the earth condensed, most of the volatile gases in the dust cloud escaped. The water that is now in the oceans and the gases in the atmosphere were not volatile at that time; they were expelled from the interior of the earth at a later date. Soon after the earth had formed, its interior began to warm up, for large amounts of energy must have been released by the decay of radioactive elements. Any slightly volatile material in the interior of the earth was, thus, forced to the surface as volcanic gas. Later, as the temperature of the interior rose even higher, nonvolatile materials were destroyed by the intense heat and the products of their destruction were forced to the surface and added as simple gases to the atmosphere.

These conclusions are very important when one considers the origins of life, for they show that even if, as is likely, complex organic substances were present in the dust cloud, they would have been largely destroyed in the period immediately after the formation of the earth. Most of the organic compounds needed for the beginning of life must have been synthesized afresh in the earth's atmosphere and oceans.

The Primitive Atmosphere

One of the most important and controversial discussions connected with the origins of life is concerned with the composition of the primitive atmosphere. To make clear the significance of this question, we must first explain what is meant by the terms "oxidizing" and "reducing." The following definitions are of limited application.

Most chemical elements form compounds with hydrogen and with oxygen. The compound that contains the highest proportion of oxygen is said to be fully oxidized, whereas the compound that contains the highest proportion of hydrogen is said to be fully reduced. The fully oxidized forms of carbon, nitrogen, and sulfur, for example, are carbon dioxide (CO_2), nitrogen pentoxide (N_2O_5), and sulfur trioxide (SO_3), respectively. The corresponding fully reduced molecules are methane (CH_4), ammonia (NH_3), and hydrogen sulfide (H_2S).

Molecules which contain less oxygen than the fully oxidized form or less hydrogen than the fully reduced form are said to be in an intermediate state of oxidation. Carbon, for example, forms a series of molecules in intermediate states of oxidation, namely carbon monoxide (CO), formic acid (HCO_2H), formaldehyde (CH_2O), and methanol (CH_3OH). This series illustrates an important point: the oxidation state can be lowered by removing oxygen or by adding hydrogen, for molecules whose compositions differ by one or more water molecules are in the same oxidation state. All biologically important organic molecules are in intermediate oxidation states.

The term "oxidizing" is used to describe conditions which tend to convert elements or compounds into their

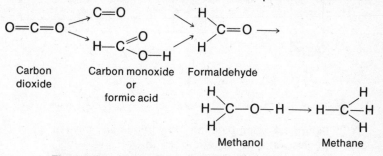

Carbon dioxide   Carbon monoxide or formic acid   Formaldehyde   Methanol   Methane

Figure 6.3. Successive stages in the reduction of carbon dioxide. Note that carbon monoxide and formic acid differ by one H_2O molecule and so are equally reduced.

more highly oxidized forms. The simplest oxidizing gas mixtures are those that contain free oxygen. In a similar way, reducing conditions are those which tend to convert compounds to more highly reduced forms, and the simplest reducing atmosphere contains free hydrogen.

Consider an atmosphere which contains only hydrogen (H_2), oxygen (O_2), and water (H_2O). Under the influence of ultraviolet light or electric discharges, free hydrogen and free oxygen combine until either the hydrogen or the oxygen is used up.

$$2H_2 + O_2 \longrightarrow 2H_2O$$

If hydrogen is initially present in excess, the final atmosphere contains hydrogen and water, whereas if oxygen is initially present in excess, the final atmosphere consists of oxygen and water. As we have seen, a simple atmosphere of this kind is called "reducing" in the former case and "oxidizing" in the latter. The word "neutral" will be used to describe a balanced atmosphere containing water, but no excess of free hydrogen or of free oxygen.

If a small amount of an organic compound is introduced into an oxidizing atmosphere containing free oxygen, it is sooner or later converted into carbon dioxide. If a small

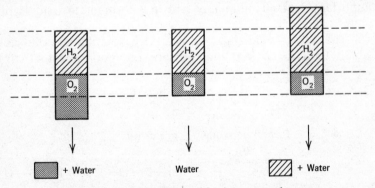

Figure 6.4. Combination of hydrogen and oxygen to give water. If the ratio of H_2 to O_2 is just 2 to 1, only water is left. Otherwise an excess of either H_2 or O_2 may remain after the maximum amount of water has been formed.

quantity of carbon dioxide is introduced into a reducing atmosphere containing hydrogen, it is sooner or later converted into methane. If a large quantity of carbon dioxide is introduced into a reducing atmosphere, the excess hydrogen may be used up before all of the carbon dioxide is reduced to methane. In this case, compounds in intermediate oxidation states, including organic compounds, may be formed. The term "neutral" will be used more generally than before to describe an atmosphere containing carbon dioxide and water, but no oxygen and no carbon compounds more reduced than CO_2. Organic compounds are formed only in an atmosphere that contains more hydrogen or less oxygen than does a neutral atmosphere.

In a similar way, if nitrogen is introduced into a reducing atmosphere, it is partially converted into ammonia (NH_3). In an oxidizing atmosphere molecular nitrogen (N_2) and oxides of nitrogen are present. A neutral atmosphere is defined as one that contains N_2 and water, but neither ammonia nor oxides of nitrogen.

The oxidizing or reducing character of the primitive atmosphere is very important for any discussion of the origins of life, because the organic compounds from which all living things are made are not stable in an oxidizing atmosphere. Furthermore, as will be seen in Chapter 7, the synthesis of important biochemical compounds takes place readily in a reducing atmosphere, but not in a neutral or an oxidizing atmosphere. Thus, most modern theories of the origins of life assume that the primitive atmosphere was reducing. It is important to see whether this assumption is consistent with the geological and geochemical evidence.

Today the atmosphere of the earth is strongly oxidizing. It contains about 80% of molecular nitrogen (N_2) and about 20% of free oxygen (O_2). The small amount of carbon in the atmosphere is present as carbon dioxide (CO_2). The chemical characteristics of the atmosphere are almost entirely determined by the free oxygen that is present in it. No organic compounds are synthesized in our atmosphere, nor are they stable if they are introduced into it. The volcanic gases from which the atmosphere was formed must have contained water and probably carbon dioxide as well. However, geologists have shown that the composition of the earth is such that little, if any, free oxygen could have been present. The oxygen which is now found in the atmosphere must, therefore, have been formed by the decomposition of water; there is no other abundant source from which it could have come.

Two mechanisms are known that could account for the formation of oxygen from water on the primitive earth. The first is the action of high-energy ultraviolet light from the sun on water molecules in the upper atmosphere. This produces hydrogen in addition to oxygen, but the hydrogen escapes from the earth's gravitational field leaving free oxygen behind. When the amount of oxygen that could have accumulated in 4.5 billion years through the photochemical decomposition of water is calculated, it is found to be less than the amount now present in the atmosphere. Thus, although some oxygen has undoubtedly been formed in this way, much of it must have had a different origin.*

The other process that produces free oxygen is photosynthesis (Chapter 4). At the present time, algae and land plants are producing vast amounts of oxygen. Since fossil algae that look very similar to modern oxygen-producing photosynthetic algae are found in Pre-Cambrian rocks, photochemical production of oxygen, has probably been going on for a very long time. It seems likely that most of the oxygen in our atmosphere has been produced by photosynthesis.

While it is generally agreed that the primitive atmosphere did not contain more than a trace of free oxygen, widely differing views have been expressed concerning its reducing character. According to one extreme view, the atmosphere was made up of completely reduced compounds—methane (CH_4), ammonia (NH_3), and water (H_2O), and probably some molecular hydrogen. According to the other extreme view, the earth's early atmosphere was neutral and contained carbon dioxide, nitrogen, and water. It is not possible at present to decide where, between these two extremes, the truth lies.

Authors who believe that the primitive atmosphere was fully reduced argue that since hydrogen is by far the most

_____
* This conclusion is now disputed.

abundant element in interstellar space, it certainly predominated in the dust cloud from which the solar system was formed. They conclude that it also predominated in the primitive atmosphere. They draw attention to the large amounts of methane in the atmospheres of the outer planets and claim that this also is evidence that large amounts of methane must once have been present on the earth. However, since the atmosphere came from inside the earth, both of these arguments are weak. Hydrogen from the dust cloud may have escaped completely as the earth was forming; the gases expelled from the interior of the earth may have been quite different from the gases making up the primary atmospheres of the outer planets.

The arguments in favor of an oxidizing atmosphere are no more conclusive. The gases evolved from volcanoes are not strongly reducing, and it is clear that an atmosphere derived from contemporary volcanic gases would contain a great deal of carbon dioxide. However, most volcanic gases are formed by the recirculation of water and gases from the earth's crust and do not come from the deeper layers of the earth. Thus, the composition of contemporary volcanic gases tells us little about the composition of the gases which come from the earth's core. Even if we knew the composition of such gases, we could not assume that it is the same as that of volcanic gases on the primitive earth, because great changes have occurred in the interior of the earth during the 4.5 billion years that have elapsed since its formation.

There is some geological evidence of a different kind which favors the view that the early atmosphere was reducing, or at least not oxidizing. Mineral deposits formed under reducing conditions contain iron in what is called the ferrous state, while mineral deposits formed when free oxygen is available always contain ferric iron. Very large amounts of ferrous iron were deposited in early Pre-Cambrian times. This shows that the early Pre-Cambrian atmosphere did not contain much free oxygen, but unfortunately it does not distinguish unambiguously between a neutral and a reducing atmosphere.

We shall see that the success of most prebiotic

syntheses which start with simple gases requires that the reaction mixtures be reducing rather than neutral or oxidizing, but it does not depend on the detailed composition of the mixture. Fully reducing mixtures containing methane, ammonia, and hydrogen or partially reducing mixtures containing carbon monoxide, nitrogen, and hydrogen behave in very similar ways. Although the nature of the earth's first atmosphere is not yet known in detail, it is almost certain that it was sufficiently reducing to make possible the organic syntheses which will be discussed in later chapters. Of course, the finer details of the composition of the earth's primitive atmosphere are a subject of great intrinsic interest, but they are perhaps less critical for theories of the origins of life than is sometimes supposed. To explain the formation of organic molecules it must be supposed that the atmosphere was reducing and not much more needs to be assumed about its composition.

We know very little about the time that it took to form the atmosphere and oceans. Presumably, water, carbon compounds, and nitrogen or ammonia were released together from the interior of the earth. After a while the atmosphere became saturated with water vapor and rain began to fall. At first small lakes must have formed and then, as more and more water entered the atmosphere, these lakes must have been enlarged to form the oceans. The accumulation of salts in the oceans depended on the weathering of rocks and may have been quite slow.

All this is important because it raises an interesting question about the environment in which life began. If the formation of the oceans was rapid, one is justified in assuming that life began in oceans and lakes similar to those with which we are familiar. If, on the other hand, the oceans formed slowly, life may have begun before the oceans reached their present size and before they acquired their present content of salts. It is not possible to decide between these alternatives. In the rest of this book we shall make the conservative assumption that, when life began, the distribution of water on the earth was much as it is now. If the oceans had been smaller and less saline than they are now, some of the difficulties connected with the accumulation of organic materials would have been less severe.

Sources of Energy

7

Energy Sources

It is generally believed that the very first stage in the origins of life was the accumulation of large amounts of dissolved organic material in the oceans and lakes of the primitive earth. The solution that was formed in this way is often referred to, picturesquely, as the prebiotic soup. It is important to know whether, when life began, the prebiotic soup was hot or cold, thick or thin, and so on. In this chapter the formation of the ingredients of the soup will be discussed in a preliminary way.

A reducing atmosphere of the kind that is thought to have been present on the primitive Earth is stable indefinitely unless it is acted upon by a source of energy. No new organic compounds are formed when a mixture of methane, nitrogen, and water, for example, is left to stand in the dark. Complex organic compounds are formed if and only if the mixture is heated strongly, irradiated with ultraviolet light, acted upon by an electric discharge, or subjected to the action of some other form of energy.

The chemistry involved in the formation of organic molecules from simple gas mixtures is too complicated to be discussed extensively here. Instead a single reaction will be described: the formation of hydrogen cyanide (HCN) from nitrogen (N_2) and methane (CH_4). This is one of the most important prebiotic reactions and, fortunately, one of the simplest.

When an electric discharge is passed through an atmosphere containing molecular nitrogen (N_2), some of the molecules absorb so much energy that they dissociate into a pair of atoms.

$$N_2 \longrightarrow 2N$$

Now, while nitrogen molecules are quite unreactive, nitrogen atoms attack almost any other atom or molecule. In particular, nitrogen atoms react with methane molecules to give hydrogen cyanide and hydrogen.

$$N + CH_4 \longrightarrow HCN + \tfrac{3}{2}H_2$$
$$\text{Methane} \quad \text{Hydrogen}$$
$$\text{cyanide}$$

It follows that when an electric discharge is passed through a mixture of nitrogen and methane, the nitrogen atoms formed in the discharge react with methane molecules in their environment to form hydrogen cyanide. In a similar way, nitrogen molecules are dissociated into atoms when nitrogen is heated very strongly or irradiated with high-energy ultraviolet light. Thus, hydrogen cyanide could equally well have been formed in the primitive atmosphere from nitrogen and methane by the action of lightning, volcanoes, or ultraviolet light.

Hydrogen cyanide, unlike nitrogen and methane, is very soluble in water. Any hydrogen cyanide formed in the primitive atmosphere would, therefore, have dissolved in raindrops and then found its way into the oceans. In the next chapter, the way in which important organic compounds are formed from aqueous solutions of hydrogen cyanide will be discussed. Here we need recognize only the important general principle that reactive molecules, like hydrogen cyanide, were formed in the primitive atmosphere. Subsequently, they dissolved in oceans and lakes, where they reacted with one another, spontaneously, to form the prebiotic soup.

It is clearly important to decide which forms of energy contributed most to the synthesis of organic compounds on the primitive earth. This will not be an easy matter until one can be certain about the composition of the primitive atmosphere. In the meantime, one can make some informed guesses.

The sun supplies far more energy to the earth than does any other source. Each year about 260,000 calories of radiant energy are incident on each square centimeter of the atmosphere. This amount of energy is sufficient to boil away a layer of water twelve feet deep covering the whole surface of the earth. However, not all of this energy is useful because light can bring about chemical reactions only if it is absorbed by one of the components of the reaction mixture; light that passes through a mixture of gases without being absorbed can have no influence on the gases.

Table 7.1. Present Sources of Energy Averaged over the Earth

Source	Energy (cal $cm^{-2}yr^{-1}$)
Total radiation from sun	260,000
Ultraviolet light beyond	
3000 Å	3,400
2500 Å	563
2000 Å	41
1500 Å	1.7
Electric discharges	4
Shock waves	1.1
Radioactivity (to 1.0 km depth)	0.8
Volcanoes	0.13
Cosmic rays	0.0015

Figure 7.1 illustrates the way in which the energy reaching the atmosphere from the sun is distributed over the

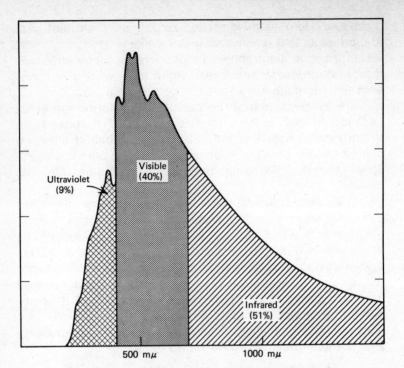

Figure 7.1. Solar radiation intensity above the atmosphere at earth's distance from the sun. In all an average of about 260,000 calories is incident on each square centimeter of the atmosphere per year. (Reproduced with permission from *Source Book on the Space Sciences* by Samuel Glasstone, Van Nostrand, Inc., New York 1965.)

infrared, visible, and ultraviolet regions of the spectrum. Most of the energy is in the visible region; the energy available falls off rapidly in the ultraviolet. Since the gases that were present in the primitive atmosphere do not absorb visible light, only the fraction of the energy present as ultraviolet radiation could have been used for the synthesis of organic compounds. The precise amount would have depended on the composition of the primitive atmosphere.

An atmosphere containing methane, ammonia, nitrogen, and water of the kind that might have existed on the primitive earth could have absorbed at most 40 calories per square centimeter per year, that is, less than 0.02% of the total solar energy. This would not have been enough to cause very extensive organic synthesis. However, it has been shown that if hydrogen sulfide or formaldehyde were present in sufficient quantities in the atmosphere, they could have absorbed a much larger amount of ultraviolet energy and made it available for the synthesis of organic compounds. At present we have no information about the abundances of formaldehyde and hydrogen sulfide in the primitive atmosphere.

The reader should realize that the amount of ultraviolet light reaching the surface of the earth today is very much less than the amount that reaches the top layer of the atmosphere. The surface of the earth and the lower atmosphere are protected by a layer of ozone (O_3) in the upper atmosphere. Ozone, unlike the other components of the atmosphere, absorbs ultraviolet light strongly and prevents it from reaching the earth. Otherwise, men would be subjected to very harmful doses of ultraviolet light whenever they were exposed to direct sunlight.

$$3O_2 \longrightarrow 2O_3$$
Oxygen   Ozone

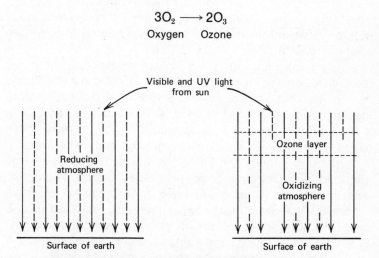

Figure 7.2. Partial shielding of the earth from ultraviolet light by the ozone layer formed in an oxidizing atmosphere.

Certainly, there was no ozone on the primitive earth because it is formed from O_2 and cannot survive in a reducing atmosphere. Many of the photochemical reactions occurring in the primitive atmosphere could not occur today near the surface of the earth because so little ultraviolet light penetrates the ozone layer.

Electric discharges have been used to study prebiotic synthesis more often than any other energy source, mainly because they produce organic compounds with high efficiency. At the present time the quantity of electic energy released on the earth is very much smaller than the total amount of solar energy incident on the atmosphere. However, all of the electric energy is in a form that is effective in synthesizing organic compounds from a reducing atmosphere, while most of the sun's energy would have passed through such an atmosphere without causing chemical change.

We do not know how often thunderstorms occurred in the reducing atmosphere of the primitive earth. Consequently, we do not know how much electric energy was available. I shall assume that the amount was not very different from that available today. However, this may be an underestimate since there are reasons for believing that thunderstorms would be very frequent in a reducing atmosphere containing ammonia.

Very large amounts of energy were released within the earth through the decay of radioactive substances. We do not believe that this energy was important for the origins of life since most of it was liberated far below the surface. Even if organic compounds were formed, they could never have found their way into the primitive oceans and lakes.

Volcanic activity, which is due indirectly to the energy liberated by radioactive decay processes, occurs at the surface of the earth and could have contributed to prebiotic synthesis. The amount of energy presently released in volcanoes is quite small, but on the primitive earth it could easily have been as much as ten times greater. Nonetheless, it is doubtful that volcanoes were the major site of organic synthesis, because no synthetic processes are known which make efficient use of volcanic energy.

Molten lava is hot enough to bring about the synthesis

of organic compounds from a reducing atmosphere, but once the lava comes into contact with the atmosphere, its surface cools off very rapidly and solidifies. Once this happens, the synthesis of organic compounds becomes inefficient, since only the gases which penetrate through the solid surface of the lava to the hotter material beneath are heated strongly.

A great deal has been written about the importance of volcanoes and hot springs for the origins of life, and it would be rash, in the absence of further evidence, to dismiss volcanic energy as unimportant. Some organic compounds would certainly have been formed at locations where a reducing atmosphere made contact with hot lava. A few realistic experiments carried out in an active volcano might answer a number of important questions—and provide a good deal of excitement for the investigators in the process.

Many other forms of energy have been proposed as causes of prebiotic synthesis, for example, cosmic rays, sonic energy generated by ocean waves, and shock waves generated by thunderstorms or by meteorites entering the earth's atmosphere. In all cases the amounts of energy involved are quite small, and only in the case of shock waves is synthesis claimed to be efficient enough to make up for this disadvantage. If the rather surprising claims for the efficiency of synthesis in shock waves can be confirmed, we will have to consider them as a potentially important source of prebiotic organic compounds.

It is unlikely that all of the organic compounds synthesized on the primitive earth were formed in the same way. We need to determine how much each energy source contributed to the synthesis of each class of compound, rather than to find the unique energy source that was responsible for all prebiotic syntheses. In my opinion, electric discharges and ultraviolet light must certainly have been important, but the role of other energy sources is as yet unclear.

It is perhaps worth emphasizing that few nonbiological organic compounds are accumulating on the earth today. Nonbiological synthesis of organic compounds does not occur in our atmosphere because so much free oxygen is present. Even those compounds that are introduced into the atmosphere as industrial pollutants are largely degraded by microorganisms. Compounds of the kind that are believed to have been formed on the primitive earth are almost all bio-degradable; the novel products of the chemical laboratory, for example, the chlorinated hydrocarbons, are the most important ones that escape degradation.

Rates of Accumulation of Organic Compounds

We must now try to get some idea about the rates at which organic compounds were formed in the atmosphere, and the amounts that accumulated on the surface of the earth and in the oceans. First, it must be realized that the time available for synthesis and accumulation of organic compounds was very long; so long, in fact, that it falls outside the range for which we have any intuitive feeling. It seems almost certain that accumulation took more than a million years, and perhaps it took closer to a billion years.

Let us suppose, for example, that the amount of organic material synthesized on the earth each year was very small, say a kilogram per square kilometer, that is about an ounce for every ten acres. Then, if all of the material formed in a billion years accumulated evenly over the surface of the earth, a layer of organic solids three feet deep would have resulted. This example should make it clear that rapid and unreasonable rates of synthesis need not be assumed to justify the existence of a rich prebiotic soup.

In fact the rate of synthesis was probably much greater than has been assumed in the above example. If electric energy was available in amounts comparable with those released on the earth today, and if this energy was used as efficiently as it is in modern laboratory experiments, about 1 mg of organic material would have been formed each year for each square centimeter of the earth's surface. This is equivalent to ten thousand kilograms per square kilometer and would lead to the accumulation of a layer of organic material three feet deep in as little as a hundred thousand years.

It seems unlikely that the efficiency of prebiotic synthesis could ever have approached that of laboratory synthesis, but even if it had been only 1% as efficient, it would have produced vast amounts of material within a period of a few million years. Similar conclusions are reached when corresponding calculations are carried out for organic synthesis initiated by ultraviolet light from the sun, if we suppose that hydrogen sulfide or formaldehyde was present in the atmosphere.

The estimates which we have made of the amounts of organic material accumulating on the primitive earth are based on the assumption that all material formed in the atmosphere would have accumulated at the surface. We know that this is an oversimplification, because many of the organic compounds formed under prebiotic conditions are unstable and decompose in much less than a million years. However, even when this is taken into account, there is every reason to believe that the oceans and lakes of the primitive earth contained abundant organic material. The prebiotic soup may easily have contained as much as a gram of organic material per liter of water. This is about a third as concentrated as an average chicken bouillon.*

* The calculation was performed for Knorr's Chicken Bouillon made up according to the supplier's instructions.

Prebiotic Synthesis

8

Introduction

The term "prebiotic chemistry" is commonly used to describe chemical reactions that are carried out in the laboratory with the intention of simulating processes that occurred on the primitive earth. The reactions themselves are often little different from those studied by chemists whose motivation is quite different. Like the man who discovered to his surprise that he had been writing prose for years, many organic chemists have been doing prebiotic chemistry for years, without realizing it.

One of the most important experiments in prebiotic chemistry was carried out in 1832 by the German chemist, Friedrich Wöhler. When Wöhler published his work, scientists still believed that there was an essential difference between the chemistry of lifeless materials, such as minerals or rocks, and the chemistry of the constituents of living organisms. The two subjects were called inorganic and organic chemistry, respectively. These terms are still used, although today the subject matter of organic chemistry includes the properties of almost all compounds of carbon, whether they are of biological origin or not.

Wöhler showed that when an inorganic compound, ammonium cyanate, is heated, an organic compound, urea, is formed. Wöhler's experiment is important because it is part of a chain of evidence demonstrating that there is really no unbridgeable gap between the chemistry of lifeless and living matter. It is interesting that Wöhler's synthesis of urea can also be considered as an important prebiotic reaction in the modern sense, since it is likely that urea was formed from ammonia and cyanate on the primitive earth.

Many reactions carried out in the late nineteenth and early twentieth centuries qualify as prebiotic. Glycine, an important amino acid, was obtained from hydrogen cyanide, for example, and sugars were synthesized from formaldehyde. Since both hydrogen cyanide and formaldehyde had already been obtained from inorganic sources, these reactions could justifiably have been described as prebiotic.

Prebiotic chemistry in the sense of the deliberate simulation of reactions that occurred on the primitive earth is of much more recent origin. Oparin and Haldane emphasized that in the period before the first living organisms evolved, the earth's atmosphere must have been reducing. They suggested that a mixture of organic compounds was formed in such an atmosphere and that the first living organisms were assembled from a selection of these compounds. We have argued that the results of astronomical and geophysical research are broadly consistent with the hypothesis that the primitive atmosphere was reducing, although we must admit that many questions of detail remain to be decided.

Synthesis of Amino Acids, Sugars, and Nucleotide Bases

In 1953 Stanley Miller, then a student of Harold Urey at the University of Chicago, subjected a mixture of methane, ammonia, hydrogen, and water to the action of an electric discharge. He was able to show that, just as Oparin and Haldane had predicted, a mixture of organic compounds including amino acids was formed. These experiments stimulated many related studies and can reasonably be considered to mark the beginning of modern work on prebiotic chemistry.

Prebiotic chemistry is a complicated branch of organic chemistry. The detailed results are of a highly technical character and can be understood only in the framework of mechanistic organic chemistry. In this chapter we will summarize the progress achieved to date without going into detail. Readers who have the necessary background in organic chemistry will find the details in the references cited at the end of the book.

The equipment used by Miller in his first experiment is shown in Figure 8.1. The small flask was filled with water, and the rest of the apparatus with a mixture of methane, hydrogen, and ammonia. The gas mixture, together with some water vapor, was caused to circulate past the tungsten electrodes by boiling the liquid water in the small flask. Then a spark discharge was formed between the electrodes by applying a high electric potential across them.

The products formed in the electric discharge dissolved in the water which liquified in the condenser and were then carried down into the small flask. In an apparatus of this kind, volatile products are constantly distilled out of the small flask along with steam and then subjected again to the action of the discharge; nonvolatile products accumulate in the small flask.

Miller's experiment is instructive because it illustrates a number of features common to many attempts to simulate the chemistry that occurred on the primitive earth. Miller chose to work with a mixture of methane, hydrogen, ammonia, and water because, at the time, it seemed probable that the primitive atmosphere contained these gases as its main components. The boiling water in the small flask was supposed to represent the primitive ocean. Finally, the electric discharge which was passed through the gas was considered as an equivalent to lightning in the primitive atmosphere. In interpreting the results, it was supposed that the

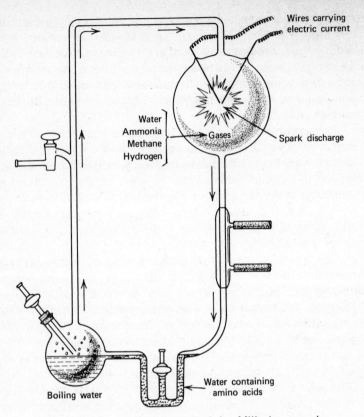

Figure 8.1. Apparatus used in Miller's experiments on the formation of amino acids from a reducing gas mixture. Reproduced with permission from *Stars, Planets and Life* by Robert Jastrow, William Heinemann, Ltd., London, 1967, p. 134.

material that accumulated in the boiling water corresponded to organic substances that would have accumulated in the primitive ocean.

Some features of Miller's experiments are clearly unrealistic: the oceans, for example, were not boiling at the time when life evolved. Miller chose to work with boiling water rather than a cold "ocean" for technical reasons; otherwise the gases in the apparatus would not have circulated quickly enough. This procedure had another important advantage, for it speeded up a number of reactions that would have been very slow in a cold solution. The effective life of a graduate student is at most a few years, so it is often necessary to achieve within a few days what must have taken hundreds of years or more on the primitive earth. That is why experiments in prebiotic chemistry rarely take the form of an attempt to reproduce exactly the conditions that existed on the primitive earth. More often extrapolation is used where exact simulation would be impractical; we study chemical reaction under conditions which allow us to work out in a short time what would have happened in a much longer time on the primitive earth.

Two types of extrapolation are particularly important. In general, the rates of chemical reactions increase as the temperature increases. It is possible to make accurate estimates of reaction rates at low temperatures if the rates are measured at two or more higher temperatures. In this way, it is possible to learn in days or weeks about reactions which would have taken millions of years in the primitive ocean.

The rates of many chemical reactions also depend in a fairly simple way on the concentrations of the reactants. This makes it possible to determine how dilute solutions in the primitive ocean would have behaved, by studying more concentrated solutions in the laboratory. This is important, because it is difficult to work with more than a few liters of solutions in an ordinary chemical laboratory, while the volume of the oceans was enormously larger.

We must now return to the results of Miller's experiment. At the end of a week, the spark was turned off and the contents of the apparatus were allowed to cool down. The solution in the small flask was then subjected to a detailed analysis. The results were extremely surprising. As much as 15% of the carbon which had originally been in the "atmosphere"; was present as identified organic compounds in the "ocean." About 5% of the carbon had been converted into important biochemical compounds. The most striking feature of Miller's results was the discovery that several of the natural amino acids were formed in substantial amounts.

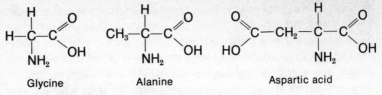

Figure 8.2. Three of the amino acids formed in Miller's experiments.

Glycine, alanine, aspartic acid, and glutamic acid were identified with certainty.

To understand how surprising this is, one must consider the results against the background of general organic chemistry. Many millions of organic compounds are known, and many thousands have structures no more complicated than those of the amino acids. Before these experiments were reported, most chemists would have anticipated that an electric discharge acting on a mixture of methane, ammonia, hydrogen, and water would produce a mixture containing small quantities of many different substances. It surely cannot be a coincidence that relatively few substances were obtained and that many of them are important biochemical compounds found in all living organisms.

These observations are most easily explained if it is supposed that the most primitive organisms were composed of organic compounds that had been formed by the action of an electric discharge on a reducing atmosphere. Since modern organisms must have retained many features of their primitive ancestors, we could then understand why cells contain so many of the compounds that Miller identified among his products.

In fact the situation is more complicated. Miller's work stimulated many further efforts to understand the organic chemistry that occurred on the primitive earth. It was soon discovered that similar sets of organic molecules, including amino acids, are formed whenever a reducing mixture of gases is heated strongly enough, irradiated with ultraviolet light or subjected to any type of electric discharge. Furthermore, it was found that the nature of the gas mixture is not critical, provided the mixture is reducing. Miller's mixture of methane, hydrogen, nitrogen, ammonia, and water can be replaced by a mixture of carbon monoxide, hydrogen, nitrogen, and water, for example. It may be concluded that amino acids were available on the primitive earth since they are formed whenever a reducing gas mixture is treated violently enough. It is not known which energy source was most important on the earth, although I personally believe that electric discharges must have made a major contribution to prebiotic synthesis.

During the twenty years that have elapsed since the publication of Miller's original paper, nearly all of the naturally occurring amino acids have been identified as products in one or the other of the many prebiotic synthesis that have been reported. The reactions are not all equally convincing, but there is little reason to doubt that if the atmosphere was reducing, many of the twenty natural amino acids would have accumulated on the primitive earth.

So far it has proved impossible to obtain amino acids from an atmosphere which contains free oxygen or from a mixture of carbon dioxide, nitrogen, and water. This is a very important negative result, for it argues strongly in favor of the Oparin-Haldane hypothesis that the primitive atmosphere was reducing. Life could not have got started in an atmosphere of the type that exists today.

This situation may seem somewhat paradoxical, since today all higher and most other forms of life are completely dependent on oxygen. Only a few types of bacteria and certain other microorganisms can survive under the conditions that were necessary for the origin of life. This is an impressive example of adaptation. Most organisms presumably began to use oxygen once it was abundant and after a time became absolutely dependent on it. Now, they can no longer survive in an oxygen-free enviroment.

When an organic chemist makes a surprising discovery, his first instinct is usually to prove that it is not surprising at all. He tries to show that the new facts that he has discovered fit in with, and perhaps extend, generally accepted theories. Miller was soon able to demonstrate that some of the amino acids among his products had been formed by a well-known route. One of the first syntheses of glycine to be reported involved heating together hydrogen cyanide and formaldehyde in an aqueous solution of ammonia. Miller showed that hydrogen cyanide and formaldehyde were formed by the electric discharge and reacted together in aqueous solution to give glycine. Thus, Miller was repeating a well-known synthesis of glycine, but generating the starting materials with an electric discharge under prebiotic conditions.

This work focused attention on hydrogen cyanide and led to the next major development in prebiotic chemistry. Juan Oro was checking whether amino acids could be formed from hydrogen cyanide and ammonia, in the absence of formaldehyde or similar substances. He found that if he warmed ammonia and hydrogen cyanide together in aqueous solution for a few days, amino acids similar to those discovered by Miller were formed. He found, in addition, that adenine was obtained in about 0.5% yield. More recently, a commercial synthesis of adenine from ammonia and hydrogen cyanide has been developed in Japan.

Adenine is, of course, one of the four bases present in RNA and DNA. It is also a component of ATP, the intermediate involved in the storage and utilization of chemical energy in biological systems. The structure of adenine is quite complicated, so much so that it is quite astonishing that it can be formed in large amounts from hydrogen cyanide in so simple a reaction. It is hard to avoid the conclusion that adenine occupies a central position in biochemistry because it is one of the few organic compounds of this degree of complexity that formed in large amounts on the primitive earth.

The evidence that many important biochemicals are formed with surprising ease under prebiotic conditions has

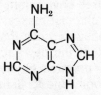

Adenine

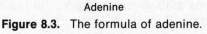

Figure 8.3. The formula of adenine.

$$H - C \equiv N \qquad H - C \equiv C - C \equiv N$$

Hydrogen cyanide   Cyanoacetylene

Figure 8.4. The formulas of hydrogen cyanide and the related molecule, cyanoacetylene.

continued to accumulate. By now, syntheses of almost all of the monomeric components of the genetic apparatus have been achieved under prebiotic conditions. Here, the description of one more example must suffice. Since four nucleotide bases occur together in RNA, it seemed likely that they were formed together by similar mechanisms on the primitive earth. It was natural, therefore, to look for a synthesis of cytosine (C) and uracil (U) from an intermediate similar to hydrogen cyanide.

An examination of the structure of cytosine (Figure 3.7) suggested that its synthesis could be achieved by starting with cyanoacetylene, a gas closely related to hydrogen cyanide. This indeed proved to be the case. However, it was surprising to find that cyanoacetylene is an abundant product in almost all prebiotic reactions that produce hydrogen cyanide. When methane and nitrogen are subjected to an electric discharge, for example, hydrogen cyanide is the main nitrogen-containing product but cyanoacetylene is also formed in reasonable amounts.

So far, nothing has been said about the prebiotic synthesis of two very different components of the genetic system, the sugars—deoxyribose and ribose. In fact, the prebiotic synthesis of ribose was achieved in the nineteenth century by a Russian scientist, Butlerov. In the Butlerov reaction, formaldehyde is shaken with chalk or lime. This very simple procedure gives a complex mixture of sugars, including ribose. Deoxyribose can be obtained by a modification of this procedure.

We may summarize this aspect of prebiotic chemistry as follows. Whenever a reducing mixture of gases containing carbon, hydrogen, nitrogen, and oxygen is treated violently enough, a similar pattern of small, highly reactive molecules is formed. These include intermediates, such as formaldehyde, hydrogen cyanide, and cyanoacetylene. In the presence of water and ammonia, the reactive intermediates

131
Prebiotic Synthesis

combine together to form more complicated organic molecules. The products are remarkable in that they include many more of the compounds important in modern biochemistry than can be attributed to chance. To explain this finding it must be supposed that the first organisms evolved from organic compounds formed by the action of high-energy sources on a reducing atmosphere and that modern organisms have evolved from primitive ones without much change in chemical composition.

It will be seen in Chapter 14 that new evidence supporting these conclusions has come from an unexpected source. Astronomers have found that interstellar dust clouds contain large amounts of organic material. The first organic compounds to be identified were formaldehyde, hydrogen cyanide, and cyanoacetylene, three of the most popular prebiotic intermediates. To some extent this must be a coincidence, but it probably indicates that our ideas on prebiotic syntheses are broadly correct. There cannot be too many radically different families of organic molecules that are formed from inorganic starting materials under prebiotic conditions.

The Formation of Polymers

9

Concentration Mechanisms

In the last chapter we saw that highly reactive organic molecules, once they had formed in the primitive atmosphere, dissolved in oceans and lakes, and that they reacted there to form amino acids and other compounds important for biochemistry. No further progress toward the appearance of life was possible until these molecules had combined to form polymers. We can be sure of this, because the operation of the genetic apparatus depends on properties of polymers that have no counterpart among the properties of smaller molecules. This is particularly clear in the case of the nucleic acids. The sequence of nucleotides in a nucleic acid transmits the information needed to specify a protein sequence. No small molecules could have performed this function because there is no way in which they could have carried so much information.

Laboratory experiments have shown that polypeptides and polynucleotides are not easily formed from amino acids and nucleotides in dilute solutions. Thus, the production of polymers on the primitive earth was almost certainly preceded by the formation of concentrated aqueous solutions, by the deposition of solid organic material, or by the adsorption of organic compounds on the surfaces of minerals, and so forth. The prebiotic soup had to thicken or solidify before the next stage in the origins of life could begin. Some of these same concentration mechanisms may have played a part in the syntheses of substances, such as sugars and nucleotide bases, which were discussed in the previous chapter.

Evaporation is the most familiar mechanism that brings about the concentration of dilute aqueous solutions. If the process is continued long enough, solid deposits are formed. There are many places on the earth where concentrated solutions or solids have been formed in this way. Desert lakes, such as the Salton Sea in California and the Dead Sea, are saturated with salt; inland salt-flats have often been formed by the evaporation of lakes.

The case of the Dead Sea is typical. Water has flowed into it from the surrounding mountains carrying with it salts extracted from the rocks. Since the climate is very hot and very dry, a great deal of water has evaporated, leaving the dissolved salts behind. Consequently, the water of the Dead Sea is saturated with sodium and magnesium salts, and large amounts of solid salt have been deposited in the area. The tourist visiting the caves at Sodom can inspect a pillar of solid salt that we believe to have been formed by evaporation, although a different account of its origin is given by the local inhabitants.

Solid deposits of salt are also formed on rocky seashores. Tidepools, formed by very high tides, sometimes evaporate to leave behind deposits of solid salt that are not wetted again until the next very high tide or the next rainfall. Salt crusts can also be formed by the evaporation of water that has splashed up above the limit of the tide.

If we are correct in believing that nonvolatile organic compounds were formed on the primitive earth, they must have been concentrated in lakes and tide pools, just as salts are today. Some water-soluble pollutants are concentrating in this way at the present time. It seems almost certain that concentrated solutions or solid deposits of organic material formed by evaporation played an important role in the origins of life.

Volatile materials cannot usually be concentrated by evaporation, since they are driven off first when an aqueous solution is heated. All water-soluble substances, including those which are volatile, can be concentrated from aqueous solution by freezing. Since many of the highly reactive molecules formed in a reducing atmosphere are volatile, concentration by freezing is likely to have been particularly important in the earlier stages of chemical evolution. Moreover, as we shall see, a number of late steps in the origins of life are also likely to have occurred at low temperatures.

If a dilute aqueous solution is cooled to a temperature below 0°C, ice begins to separate out. Since organic compounds are not soluble in ice, they are left behind in the solution. Thus, as more and more ice is formed, the solution left behind becomes more and more concentrated. This process continues until the solution is saturated. At this point the whole mass solidifies to give a mixture of solid organic material and ice crystals.

This method of concentrating organic solutions is put to good use in the production of applejack in Canada and the northern United States. Barrels of cider are left out for long periods in winter until most of their contents are frozen. The liquid that remains contains all of the alcohol that was originally present in the cider and almost all of the flavoring materials. This provides an advantageous method of preparing a strong, palatable liquor. Most governments collect taxes on distilled liquor, but very few, if any, tax liquor concentrated by freezing.

A number of reactions that must have been important on the primitive earth have recently been shown to occur most readily in cold or partially frozen solutions. The synthesis of adenine from hydrogen cyanide is an important example; the template reactions to be discussed in Chapter 10 are others.

Other concentration mechanisms that may have operated

The Formation of Polymers

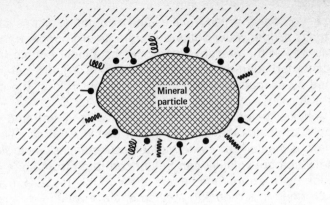

Figure 9.1. Mineral particle with adsorbed molecules.

on the primitive earth depend on adsorption. Many of the most important prebiotic molecules stick tenaciously to the surface of certain minerals. They are extracted from the bulk of an aqueous solution and concentrated in a narrow zone at the liquid–solid interface.

The clay minerals have very large surface areas and can, therefore, extract large amounts of organic material from aqueous solution. Bernal was one of the first scientists to emphasize that clay surfaces may have been the site of important prebiotic condensation reactions. Recently, as we shall see, evidence has accumulated to show that important condensation reactions do indeed take place very efficiently on the surface of a clay called Montmorillonite. Other experiments suggest that other substances including the most common phosphate minerals, the apatites, catalyze certain prebiotic condensations.

Most organic compounds dissolve in water to give true solutions, that is solutions that contain only isolated solute molecules; this is the case for glucose and acetic acid, for example. A number of compounds, detergents and soaps, for example, do not dissolve to form true solutions, but instead disperse to give small more or less spherical particles of organic material (see Chapter 4). Oparin called these colloidal droplets "coacervates", a term which is now widely used. Since many organic substances are soluble in coacer-

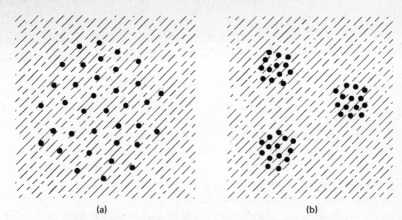

(a)   (b)

Figure 9.2. The difference between (a) a true solution and (b) a colloid. In a true solution, solute
molecules are well-separated. In a colloidal aggregate the solute molecules cluster together.

vates or are adsorbed on their surfaces, Oparin suggested
that, on the primitive earth, coacervate droplets provided the
most important site for prebiotic condensation reactions.

From many points of view this is an attractive idea.
Organic molecules dissolved in a coacervate droplet would
exist in a locally nonaqueous environment, even though the
droplet was floating in a vast excess of water. Many condensation reactions go best in the absence of water and these
would be favored if they were carried out in coacervates. So
far little experimental evidence has been produced to show
that important prebiotic condensation reactions do occur
under these conditions, but the subject is of considerable
interest and deserves further study.

Condensation Reactions

The most important biological polymerization reactions are
all dehydrations, that is reactions in which the link between
neighboring monomers is established by the removal of
water. This is particularly clear in the case of peptide-bond
formation, e.g.,

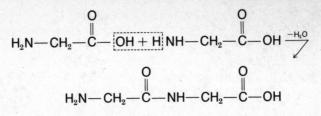

It must be admitted from the beginning that the way in which condensation reactions occurred on the primitive earth, is not understood. A few condensations are known to proceed under prebiotic conditions, but no efficient synthesis has been discovered that can plausibly be considered as a forerunner of protein synthesis or polynucleotide replication.

The most important biochemical condensations take place in a predominantly aqueous environment. It seems unlikely that any form of nucleic acid replication, whether enzymatic or not, could occur in the absence of water. This makes it almost certain that some of the earliest prebiotic condensations took place in solution. However, attempts to simulate these reactions have not been very successful; it is difficult to form a polymer by removing water from monomers if the environment already contains a vast excess of water.

It is much easier to dehydrate mixtures of amino acids or nucleotides by heating them until water is driven off. Some thermal condensations can be made to proceed quite efficiently in this way. Unfortunately, there is no obvious relationship between any of these reactions and modern biological condensations. If dry thermal reactions were important for the origins of life, it is hard to understand how they were replaced by solution reactions later in the course of biochemical evolution.

I believe that the two types of reaction occurred in close association on the primitive earth. Random polymers and reactive intermediates were formed in efficient thermal reactions. When the mixtures of products formed in this way were wetted, for example by rain or dew, they reacted further in solution. If this picture is correct, modern biological condensations evolved from these latter solution reactions.

Of course, there are other possibilities. Perhaps there are more efficient methods of condensing monomers in an aqueous environment than any discovered so far. In that case, the genetic system may have evolved without the intervention of thermal condensations. In view of these uncertainties, both types of reaction will be described.

Thermal Condensations

If a solution of amino acids is heated gently, water is driven off and a solid cake of organic material is left behind. On stronger heating, chemically bound water is eliminated and, under certain circumstances, peptides are formed. If, for example, a mixture of all twenty naturally occurring amino acids is heated, good yields of polypeptides are obtained; they have been called proteinoids, since they are claimed to resemble proteins quite closely.

From many points of view, this is an ideal prebiotic condensation. The reaction conditions are very simple and no reagents are needed other than the amino acids themselves. However, the reaction does not occur at temperatures substantially below 130°C. The highest temperatures reached at the surface of the earth today, except in volcanoes, is close to 80°C, and it seems unlikely that substantially higher temperatures occurred at the surface of the primitive earth. Thus the thermal polymerization of amino acids by direct heating could have occurred only in volcanoes.

Reasons for questioning that volcanoes were important for the origins of life have been given already. The thermal synthesis of polypeptides, if it was a significant prebiotic reaction, probably occurred at lower temperatures, perhaps in the presence of organic catalysts. However, this is an undecided issue and a number of authors, particularly the American scientist Dr. Sidney Fox, believe that volcanoes played a major role in the synthesis of organic polymers on the primitive earth.

So far, attempts to form nucleic acids by dry heating have had even less success than the corresponding attempts to synthesize proteins. Many different steps are involved in the formation of a nucleic acid from its components. Since none of them occurs at all easily on heating under plausible prebiotic conditions, it must be concluded that the synthesis of nucleic acids is unlikely to have occurred in this way on the primitive earth.

Despite these disappointments, or perhaps because of them, slow progress is being made along slightly different lines. Living organisms are able to bring about difficult syntheses with the help of enzymes. There is no possibility that enzymes, in the sense of polypeptides with precisely defined amino acid sequences, existed on the primitive earth before the appearance of life. Nonetheless, simple organic or mineral catalysts that permitted thermal condensation reactions to proceed at temperatures as low as 70–80°C, may have been present.

So little is known about catalysis in the solid state that we cannot predict which minerals or organic molecules would catalyze thermal condensations. Searching for a prebiotic catalyst is, therefore, rather like looking for a needle in a haystack. Nonetheless, a few interesting catalysts have been discovered by trial and error. It was stated in Chapter 8 that the synthesis of urea by Wöhler can be considered as the first prebiotic synthesis. It turns out that urea is an excellent catalyst for the introduction of inorganic phosphate into nucleosides, one of the difficult steps in the formation of nucleic acids from their components. With the help of urea, it is possible to add phosphate to nucleosides and to form short polynucleotides from them under conditions that occur today in many desert areas.

Many different organic compounds must have been formed on the primitive earth, and some of them certainly became concentrated in rather limited geographical regions. Areas that were rich in catalytically active material would have provided sites for later stages in the evolution of life. Sites where mineral catalysts were abundant could also have been important. The search for prebiotic catalysts among compounds that could have accumulated on the primitive earth is sure to become an increasingly important aspect of prebiotic chemistry.

Condensations in Solution

Next, attempts that have been made to synthesize polymers in aqueous solutions must be described. Here, we are faced with the difficulty that all proteins and nucleic acids are decomposed into their constituents by water; the proteins react with water to give amino acids while the nucleic acids decompose first into nucleotides and then into sugars, bases, and inorganic phosphate (see Figure 3.7). The reverse reaction, the formation of proteins or nucleic acids from monomers, never occurs spontaneously in aqueous solution.

Some of the most important aspects of modern chemistry are concerned with the concept of equilibrium (Chapter 4). A system is said to have reached chemical equilibrium when there is no longer any tendency for it to undergo further chemical change. Clearly a solution of a protein is not in equilibrium, since proteins in time decompose into amino acids. On the other hand, a solution of amino acids is in equilibrium, since there is no tendency for amino acids to react to form peptide bonds. Proteins and nucleic acids, in solution, do not reach equilibrium until they have broken down into their components.

It was shown in Chapter 4 that a chemical system can never be shifted from its equilibrium position without the expenditure of energy. It follows that energy must be supplied to the monomers if amino acids are to be converted to polypeptides, or nucleotides to polynucleotides. In living systems the energy is always provided by the breakdown of ATP. We would like to know where the energy came from on the primitive earth.

Many experimental studies, both of biological and non-biological condensation reactions in solution, have shown that a single basic mechanism is involved. Monomers always react first with the source of energy to form activated intermediates, also called high-energy intermediates. These activated intermediates, because they contain a large amount of chemical energy, react to form polymers—remember, the original monomers could not form polymers directly. The sequences of reactions involved in protein synthesis (Chapter 4) illustrates this principle.

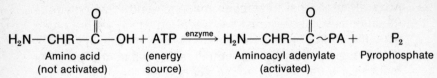

$$H_2N—CHR—\overset{\displaystyle O}{\overset{\|}{C}}—OH + ATP \xrightarrow{enzyme} H_2N—CHR—\overset{\displaystyle O}{\overset{\|}{C}}\sim PA + P_2$$

Amino acid   (energy   Aminoacyl adenylate   Pyrophosphate
(not activated)   source)   (activated)

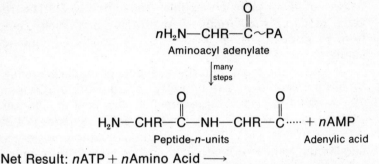

$$nH_2N—CHR—\overset{\displaystyle O}{\overset{\|}{C}}\sim PA$$

Aminoacyl adenylate

↓ many steps

$$H_2N—CHR—\overset{\displaystyle O}{\overset{\|}{C}}—NH—CHR—\overset{\displaystyle O}{\overset{\|}{C}}\cdots + nAMP$$

Peptide-n-units   Adenylic acid

Net Result: nATP + nAmino Acid ⟶
nPeptide + nAMP + nPyrophosphate

Amino acids react with the primary source of energy, ATP, to give high-energy intermediates called aminoacyl adenylates. These are the intermediates from which proteins are ultimately formed.

It seems almost certain that similar mechanisms were involved in any prebiotic condensations that took place in aqueous solution. This raises two connected questions: What were the important high-energy intermediates on the primitive earth? How were they formed? Several answers to these questions have been offered, but none of them is totally convincing.

High-energy intermediates could have formed when dry mixtures of solids were heated with suitable catalysts. One attractive scheme involves high-energy phosphate compounds from the very beginning of the evolution of the genetic apparatus. ATP and related compounds could have formed from nucleotides and inorganic phosphates in the presence of urea or some other catalyst. The high-energy phosphates formed in this way would have undergone further condensation when the products of the thermal reaction came into contact with water.

A number of special sites on the primitive earth can be imagined where a sequence of reactions of this kind could have occurred. In deserts, the surface may be heated to 80°C during the day and then wetted with dew at night. Tidepools and shallow desert lakes are locations where hot-dry and cool-wet conditions alternate at intervals of months or even years. It seems quite likely that the polymerization phase of the origins of life took place at special sites such as these. Alternatively, polymerization could just possibly have occurred near volcanoes, where certain rock surfaces are maintained at steady high temperatures except when drenched with rain.

A number of completely different schemes in which activated intermediates are formed directly in solution have been proposed. Unfortunately, they are not easily explained without going into great detail. Here it must suffice to remark that certain highly reactive molecules that could have formed in the primitive atmosphere, for example, cyanoacetylene, react with amino acids and nucleotides in aqueous solution to form high-energy intermediates. These intermediates subsequently react to form polymers. Unfortunately, the reactions are so inefficient that it has not been possible to synthesize polypeptides or polynucleotides in reasonable amounts by these methods.

More and more attention is now being given to the search for materials that catalyze condensation reactions in aqueous environments. Some of the most exciting results from such work are those being reported from Aharon Katchalsky's laboratory in Israel. He and his collaborators have shown that a common clay mineral, Montmorillonite, adsorbs certain activated amino acids and converts them in almost 100% yield to long polypeptide molecules by combining a number of mineral catalysts. Katchalsky's group succeeded in forming polypeptides in good yield directly from amino acids and ATP—this is an impressive achievement.

The particular processes discovered by Katchalsky may

or may not have been important for the origins of life. However, his work indicates clearly that polymerization of the type that must be postulated to account for the origins of life do take place on the surface of simple mineral catalysts. They may also take place on coacervates, but this remains to be demonstrated.

Replicating Molecules and Natural Selection

$$10$$

Molecules and Natural Selection

In Chapters 7–9 we discussed a formation of simple organic molecules on the primitive earth and the way in which they reacted with one another to form polymers. These reactions must have set the stage for the origins of life, but they are very different from characteristically biological processes, such as DNA replication and protein synthesis. The present chapter deals with the problem of the evolution of a self-reproducing biological system from a family of random polymers. This transition was the crucial phase in the origins of life.

The nature of the critical transition to a biological system can be explained only after we have seen how the theory of natural selection can be applied to the behavior of populations of replicating macromolecules. The principle of natural selection was originally proposed by Darwin and

Wallace to account for the way in which animals evolve. Since we shall be dealing with the evolution of molecules rather than that of animals, we shall have to discuss the theory in a somewhat unconventional way.

The theory of natural selection can be applied to the behavior of any population of reproducing entities. Here, these entities are referred to loosely as organisms, without implying that they are normal living organisms; from this point of view, an organism could be an elephant, but it could equally well be a molecule of RNA. The aim of the theory of natural selection is to account for the changes that occur in populations of organisms as they compete against each other.

To describe a population, one needs to classify its members and then to specify how many individuals there are in each of the classes. Sometimes, the way in which the characteristics of the members of a class change with time is of interest. On other occasions, only the variation in the sizes of the classes may be of interest. In considering the human inhabitants of the U.S.A., one might be concerned, for example, with the way in which the average height is changing in occidental and oriental communities. Alternatively, one might be interested only in the proportions of occidentals and orientals in the total population.

When we are interested only in the numbers of members in different well-defined classes, the law of natural selection takes a simple form: the organisms that reproduce most efficiently sooner or later dominate the population, and all other closely related organisms are eliminated. This principle had been recognized long before Darwin developed his complete theory, for example by Malthus at the end of the eighteenth century.

We have to be careful about the meaning we give to the expression "reproduce efficiently." An organism that produces few offspring all of which reach reproductive age often outgrows a competitor that produces more offspring many of which die in childhood. It is possible to take such complications into account and to derive a number that describes the overall rate of growth of a population (the Malthusian parameter). For many purposes, it is more convenient to specify the rate of growth indirectly by giving the time that it takes for a population to double in size. Natural selection guarantees that the organism that doubles the size of its population in the shortest time sooner or later eliminates all its competitors.

To see how these ideas apply to the evolution of macro-molecular systems, consider two families of self-replicating organisms, say RNA molecules, that are competing against each other. Suppose the slower-growing population takes a minute to double and the faster-growing population takes 50 seconds. Then, in 5 minutes, the two populations go through 5 or 6 doublings, respectively; the ratio of the size of the faster-growing population to that of the slower-growing population, therefore, doubles every 5 minutes. After 5 hours, a member of the faster-growing population would produce on the average $2^{60}(10^{18})$ times more descendants than a member of the slower-growing population.

When we study natural selection experimentally, we are never able to maintain very many rounds of replication before the supply of nutrients runs out. In typical experiments on bacteria or replicating RNA molecules, a small initial population is chosen and permitted to go through about ten rounds of replication. At this point it is necessary to add a small sample from the expanded population to a new supply of nutrients to start the second cycle of growth. This process may be repeated many times.

Let us return to our numerical example and see how this new procedure would work out. It will be assumed that the growth cycles last for five minutes and are always initiated with samples of 1,000 organisms. It will also be assumed that the original population contains equal numbers of fast- and slow-growing RNA molecules.

Our very first sample would contain about 500 fast-growing and 500 slow-growing organisms. We have seen that the ratio of the size of the fast–growing population to that of the slow-growing population doubles every five minutes, so our second sample would contain about 667 fast–growing and 333 slow-growing organisms. It is easy to calculate that the ratio would approximate 800:200 (4:1) at the third sampling, 889:111 (8:1) at the fourth sampling, and

Replicating Molecules and Natural Selection

so on. By the ninth sampling we would expect only about 4 slow-growing organisms to be left. Three further cycles would usually be enough to eliminate the last slow-growing organisms. At that point selection would be complete; no slow-growing organisms would be found in any subsequent cycle. It should be noted that the proportion of slow-growing organisms decreases geometrically with time; if each growth cycle were initiated with a sample containing 1,000,000 organisms instead of 1,000, selection would take only twice as long.

The amount of food available for growth is always limited under natural conditions, so that expansion of populations of organisms is always subject to restrictions similar to those discussed above. When a sample containing different kinds of organisms is introduced into a hospitable environment, the population grows until it is limited by the supply of nutrients. Then the population size remains constant while the different types of organisms in the sample compete against each other. Sooner or later, the organisms that replicate most efficiently outgrow all their competitors. If no new types of organisms were produced by mutation, selection would then be at an end.

The next part of the theory of natural selection becomes relevant when one considers the way in which new types of organisms evolve. Notice that this part of the theory presupposes that new variants can arise spontaneously. If organisms were always exactly like their ancestors, one population might displace another, but no population with novel characteristics could evolve; some species might become extinct, but no new species could evolve to replace them.

In Darwin's time little was known about the sources of the variability that make evolution possible. Darwin's own ideas on the subject proved to be incorrect. The basis for an adequate solution of the problem of variation in plants and animals was proposed by Mendel as early as 1865, but it was so much in advance of its time that it was neglected. Mendel's laws were not rediscovered until the beginning of the twentieth century.

Mendelian genetics is a complicated subject. The difficulties are due to the manner in which each parent contributes to the genetic constitution of the offspring in sexually reproducing species. We do not have to worry about any of these difficulties, because we shall be concerned with evolution at a period before the development of sexual mechanisms of reproduction. A "progeny" *molecule* is derived from a single parent and, in the absence of mutation, is indistinguishable from that parent.

Let us now consider a family of replicating nucleic acids competing against each other in a constant enviroment. (The argument would be exactly the same for any other type of replicating molecule.) In the absence of mutation, those molecules in the population that replicate fastest would eliminate all others, and after that no further change could occur. However, since nucleic acid replication is imperfect, mutant molecules would occasionally be produced that replicated at a rate different from that of the standard type. Mutants that replicated more slowly would be eliminated by selection in the usual way, but any mutant that replicated faster might take over.

We can now see how a population of replicating molecules would be likely to evolve when transferred into a *new* environment. At first, the most efficient type in the original population, F say, would take over. In a new environment it is almost certain that a better adapted mutant M_1 would soon appear and displace F. In time M_1 would, in turn, be displaced by a faster-growing mutant M_2, and so on. However, this process could not go on forever. Each successive type that dominated the population would have to be better adapted than any of its predecessors. In time a mutant M_A would appear that was as well adapted as possible; once this mutant took over, evolution would be at an end.*

When we discuss the replication of polymers on the primitive earth, we assume that the environment differed from place to place. Some regions were hot and others cold; in some the prebiotic soup was concentrated and in others it was more dilute. Under these circumstances it is likely that different replicating polymers took over the various regions. A primitive form of "speciation" must have existed from the beginning.

The reader may be surprised to realize that we have just

---

* This is an oversimplification, but adequate for the present purposes.

Replicating Molecules and Natural Selection

described, in a general way, the crucial step in the origins of life. Families of replicating molecules, competing against each other and evolving under the influence of mutation and natural selection, must have passed by imperceptible steps into the richly diversified species that we know today. In few other cases can such a complicated story have had such a simple beginning.

One may or may not agree that a family of replicating macromolecules of the type described above can be fairly described as living. However, it is certainly true that they exhibit at best a rather uninteresting form of life. The next stage in the evolution of life must have been the selection of polymers that could do something more interesting than replicate at the expense of preformed monomers.

Since at this stage, the replicating entities were no more complicated than polynucleotides, further adaptation must have depended on simple interactions between these polymers and other nonreplicating molecules in their environment. First, the polynucleotides "learned" to capture whichever small molecules in the prebiotic soup could help them to replicate faster. Later they must have learned to join together small molecules in the prebiotic soup and to hold onto the polymeric products. Gradually the nucleic acids came to assume the role of dictators, controlling the chemistry of their environment for their benefit.

The most important of the interactions between replicating nucleic acids and smaller molecules involved the amino acids; the genetic code is the final product of the evolution of nucleic acid/amino acid interaction. The invention of protein synthesis permitted nucleic acids to direct the synthesis of enzymes and in this way establish almost complete control over their chemical environment. The variety of living forms is an expression of the diverse strategies that nucleic acids have evolved in order to make use of their environment to favor their own replication.

These are the ideas that motivate much modern work on the origins of life. The point of view is deliberately limited. We concentrate on the replication of nucleic acids and on the way in which nucleic acids use their environment to facilitate their own replication. In this field, the molecular biologist is like a man who regards an elephant as a complicated device designed to replicate elephant DNA; the other point of view, that elephants are interesting animals in themselves, is temporarily forgotten. In the next section we shall concentrate on the replicating nucleic acid molecules and the adaptations which the successful ones underwent to outgrow their neighbors.

Nucleic Acids and Proteins

In the last section we discussed the application of the theory of natural selection to the evolution of families of replicating organic polymers. Although the arguments used apply equally well to the evolution of any family of polymers, it was assumed that the nucleic acids were the first genetic molecules on the primitive earth. Similarly, it was supposed that polypeptides played an important part in the early stages of the development of life. These simple and plausible assumptions need to be justified, since they have often been criticized in the past.

It has been argued that the biochemistry of the first organisms may have been quite different from contemporary biochemistry, and hence that the first genetic apparatus may not have been composed of nucleic acids and proteins. This argument is logically sound. The hypothesis that nucleic acids and proteins have been involved from the very beginning of life cannot be accepted without some justification.

Amino acids and nucleotides are formed readily under prebiotic conditions, so polypeptides and polynucleotides must have been amongst the most abundant polymers formed on the primitive earth. Furthermore, we shall see that there is evidence suggesting that nucleic acids could have replicated in the absence of enzymes. These experimental observations certainly make it seem very likely that the genetic apparatus has always been composed of proteins and nucleic acids, but they are not sufficient to prove the point.

There is an additional and very powerful argument for

working with polynucleotides and polypeptides. No detailed description of a genetic system based on polymers other than nucleic acids has ever been proposed. At the present time, studies of nonenzymatic replication of macromolecules must deal with polynucleotides, because we know of no other replicating polymers. Similarly, we know of no materials other than polypeptides that can serve as models of primitive enzymes.

Until recently, theories of the origins of life were often based on the assumption that the first organisms to evolve on the earth were made entirely of protein. This view is no longer held so widely, because it is thought unlikely that any molecule related to present-day proteins could have replicated; an organism composed entirely of proteins could not have evolved by mutation and natural selection. Polypeptides may have been important as catalysts at an early stage in the origins of life, but they could not have functioned as the first genetic material.

Polynucleotides, unlike proteins, do undergo reactions related to residue-by-residue replication under prebiotic conditions. This has been demonstrated in recent experiments that use artificial polymers related to nucleic acids. Poly U is a polymer that contains only one type of residue, uridylic acid (U), so it can be considered as a very simple nucleic acid. Poly U has the remarkable property that, when mixed with monomeric derivatives of adenylic acid (A) at sufficiently low temperatures, it organizes the adenylic acid residues into a helix. Once the A residues are present in a helix, they are more easily joined together than when they are free in solution. Thus, at low enough temperatures, poly U directs the synthesis of polymers of A. Reactions like this are called template-directed reactions, because the polymer acts as a template bringing together the monomers in a way that makes it easier to join them up.

Other experiments have shown that poly C, a simple polymer of cytidylic acid, directs the synthesis of polymers of G. On the other hand, poly U has no influence on the polymerization of derivatives of G, and poly C has no influence on the polymerization of derivatives of A. Thus the Watson-

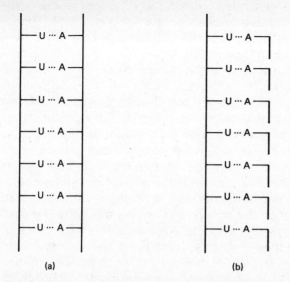

Figure 10.1. (a) A poly U-poly A double-helix; (b) part of a poly U-mono A triple-helix. The A's are arranged head-to-tail, ready to be joined up. For simplicity the second poly U chain is omitted.

Crick rules (Chapter III), that A pairs with U and G pairs with C, are obeyed in template–directed reactions. These reactions almost certainly depend on interactions between bases that are the same as those that make enzymatic DNA-replication possible (Chapter IV).

We are still far from being able to replicate a nucleic acid without the help of enzymes. Nonetheless, the information that we have already obtained is of great importance. It shows that very crude template-directed processes corresponding to nucleic acid replication could probably have occured on the primitive earth. Since template–directed reactions are successful only at low temperature, it suggests also that the first stages in the evolution of the genetic apparatus occurred in a cool prebiotic soup.

If we accept these arguments about the properties of polypeptides and polynucleotides we can clarify a subject

Replicating Molecules and Natural Selection

that has generated a good deal of heated discussion — which came first, the proteins or the nucleic acids? To answer the question we need first to distinguish carefully between two classes of polypeptides.

If the sequence of a polypeptide is determined, wholly or in part, by the sequence of a preformed polynucleotide, it will be referred to as an informed polypeptide. Otherwise a polypeptide is called noninformed.* By definition, informed polypeptides could not have existed in the absence of nucleic acids; noninformed polypeptides, on the other hand, could have accumulated before the first nucleic acids.

There is an extremely important distinction between informed and noninformed polypeptides — natural selection can operate on the former, but not on the latter. To see this we need only recollect the previous discussion. Since polypeptides cannot replicate, there is no known way in which one generation of noninformed polypeptides can influence the next. In the absence of some form of "reproduction" it is unlikely that natural selection could occur. A system containing nucleic acids and informed polypeptides is subject to natural selection; those nucleic acids that direct the formation of useful informed polypeptides are successful in eliminating their less "talented" competitors. Notice that natural selection does not act on the informed polypeptides directly, but on combined systems of informed polypeptides and "informing" polynucleotides.

We can now answer the original question quite precisely: natural selection could not act on protein sequences until nucleic acid replication was underway. On the other hand, noninformed polypeptides could well have existed from the beginning and they could possibly have catalyzed the formation and replication of the first nucleic acids.

We have no information at present on this last point. It is possible that primitive nucleic acids were formed, without the help of polypeptides, on the surface of some mineral. Alternatively, coacervates and organic catalysts, including noninformed polypeptides, may have facilitated the forma-

---

* The words random and nonrandom are confusing in this context. Noninformed polypeptides are not necessarily random.

Steps toward a Solution

tion or replication of the most primitive nucleic acids. On the whole, I find the second alternative more convincing, because it allows more gradual transition to be made from primitive to contemporary nucleic acid replication.

The Evolution of the Genetic Apparatus

From now on it will be assumed that the most primitive genetic process is polynucleotide replication and that protein synthesis is the major adaptation that permitted nucleic acids to control their chemical environment. Our next problem is to understand in more detail the sequence of steps that occured on the primitive earth and led to the evolution of the genetic apparatus in its final form.

At this point it will be useful to review some of the material presented in Chapter 3. DNA is a regular polymer composed of just four monomeric components, T, C, A and G. In all living organisms DNA functions as a master copy of the genetic material; the replication of DNA is essential for the propagation of genetic information from one generation to the next. DNA also directs the synthesis of messenger RNA; messenger RNA is an intermediary that carries the information originally coded in DNA to the protein-synthesizing apparatus.

Protein synthesis is a complicated process in which messenger RNAs determine the sequence of new protein molecules. The relation between the sequences of nucleotides in a messenger RNA and the sequence of amino acids in the protein which it specifies is determined by the genetic code. The genetic code assigns an amino acid to a sequence of three nucleotides; it is therefore called a three-letter code. The genetic code includes triplets of nucleotides that signal the termination of a protein chain as well as those that specify amino acids.

When thinking about the evolution of this complicated system, it is essential to remember that it is a product of natural selection and not the construction of a rational biochemist. It is a mistake to believe that the most primitive genetic system was as well-defined as the present system, but

Replicating Molecules and Natural Selection

simpler. On the contrary, the first genetic system is likely to have been less well-defined and possibly more complex.

It is often asked whether the first replicating nucleic acid was an RNA or a DNA. Probably it was neither; the first nucleic acid could well have contained ribonucleotides, deoxyribonucleotides, and derivatives of other sugars. Anything in the prebiotic soup that fitted into a replicating structure would have been utilized. The components of DNA and RNA, if they were both present in the prebiotic soup, could have been directed into distinct polymers only after the evolution of enzymes. The first nucleic acids may well have contained only two different bases, rather than four, but conceivably they could have contained more than four. In any case, the replication of primitive nucleic acids must have been less accurate than enzymatic DNA replication. Similarly, the earliest proteins must have contained a complex mixture of amino acids derived from the prebiotic soup. No doubt some of the amino acids in primitive polypeptides are no longer represented in proteins; conversely, some of the twenty naturally occurring amino acids may have been absent from primitive polypeptides. It is possible that compounds other than amino acids occasionally found their way into primitive proteins.

The modern protein-synthetic apparatus is made up of more than a hundred proteins and nucleic acids. The most primitive form of coding can have involved no more than a few partially ordered polymers. It must, therefore, have lacked almost all of the specificity of modern protein synthesis.

The primitive protein-synthetic apparatus, for example, cannot have discriminated between closely related amino acids. It is unlikely that leucine, isoleucine, and valine were recognized as different amino acids before the evolution of the protein-synthetic apparatus was well-advanced. The primitive genetic code, therefore, specified classes of related amino acids, rather than individual amino acids. The very first form of coding may have differentiated only two classes of amino acid. Similar arguments suggest that there could have been no efficient stop and start signs in the most primitive code; "punctuation" could not be achieved without the help of well-defined proteins. Thus, the length of primitive informed polypeptides cannot have been specified accurately.

It seems clear, therefore, that the first informed polypeptides were families of partly-ordered peptides of variable length. They may have acted as weak nonspecific catalysts, or they could have performed an even less exacting function, for example, as "glue" holding together coacervates in the prebiotic soup. It would certainly be a mistake to think of primitive informed polypeptides as though they were modern enzymes.

So far we have emphasized the sloppiness of primitive protein synthesis. Now we must deal with the one feature of protein synthesis that must have been sharply defined from the beginning. The genetic code, from a very early stage in its evolution on, must have been a three-letter code. There is no obvious reason why a two-letter or four-letter code should not have evolved in the primitive earth. However, no transition from an advanced two- or four-letter code to a three-letter code would have been possible. Such a transition would have led to a disastrous misinterpretation of all the genetic information that had been accumulated by natural selection.

We do not understand much about the later stages in the evolution of the code. Somehow, an inaccurate system must have pulled itself up by its bootstrings; polypeptide sequences that were poorly defined must have combined together to build new protein-synthetic apparatus capable of

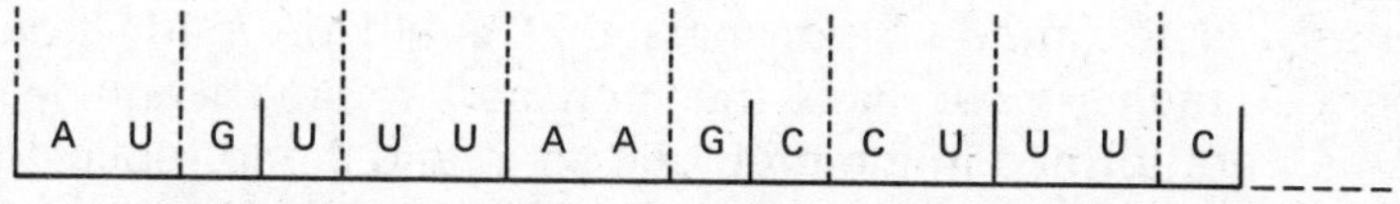

Figure 10.2. The impossibility of switching from a two-letter code to a three-letter code. After evolving to direct the synthesis of a particular sequence of 7–8 amino acids, the polynucleotide shown above would have to switch to determining a completely unrelated sequence of five amino acids. Clearly the new sequence could not function in the same way as the old one did.

157
Replicating Molecules and Natural Selection

synthesizing slightly better defined sequences. The code must gradually have improved in discrimination until it "crystallized" into its final shape.

We do not know whether the structure of the genetic code is a historical accident or not. The code may have developed in its present form because of specific interactions between amino acids and trinucleotides. Perhaps glycine interacts more strongly with the sequence GGG (or the complementary sequence CCC) than with any other trinucleotide. In that case GGG would probably code for glycine if the genetic apparatus evolved a second time. Alternatively, the relationship between trinucleotides and amino acids specified by the code could have been determined by arbitrary factors. Perhaps glycine and GGG happened to concentrate in a particular tidepool and this led to their association in the code. If so, the genetic code, if it evolved again, would be unlikely to associate GGG with glycine.*

Theories of the later stages in the evolution of the genetic code are not described here since they are all so highly speculative. Experimental work on this problem is beginning in several laboratories. Hopefully, we shall soon be able to report some progress.

Summary

One must invoke the principle of natural selection to explain how a biological system could have evolved from a prebiotic soup which contained only families of small organic molecules and the random polymers that could be made from them. It seems most likely that polymers related to nucleic acids were formed in the prebiotic soup and could replicate, even if only inaccurately, under prebiotic conditions. Natural selection would then have allowed the polymers that replicated most efficiently to win out.

The operation of natural selection would have favored those nucleic acid-like polymers that could improve their

---

* There are more plausible reasons for believing that historical accidents determined the structure of the code, but explaining them would take us too far afield.

competitive position by making use of small molecules in their environment. The amino acids were among the molecules used in this way. The genetic code is the result of an elaborate series of adaptations by means of which polynucleotides come to control their environment by directing the synthesis of structural and catalytically active polypeptides. Virtually nothing is known about the successive steps in this adaptation. This is perhaps the most challenging aspect of the problem of the origins of life.

Appendix to Chapter 10—Optical Activity

Mirror symmetry is familiar in everyday life. There is a sense in which a left-hand glove and a right-hand glove have the same shape. Nonetheless one cannot be superimposed on the other, and they interact very differently with an object such as a right hand. Pairs of objects like this are said to be related as nonsuperimposable mirror images. Many objects can be superimposed on their

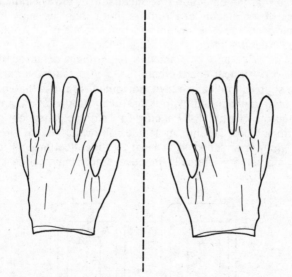

Figure 10A.1. Right- and left-hand gloves are mirror images, but no amount of reorientation permits a right-hand glove to be superimposed on a left-hand glove.

Replicating Molecules and Natural Selection

mirror images—tennis rackets, for example—but most complicated objects found in nature do not possess this poperty.

Molecules, like other objects that have a well-defined spatial structure, may or may not be superimposable on their mirror images. It will be obvious from what has been said above that large irregularly shaped molecules usually cannot be superimposed on their mirror-images. The existence of pairs of molecules having the same kind of relation to each other as a right-hand and a left-hand glove has important consequences for organic chemistry.

Pasteur discovered that certain, apparently pure, organic chemical compounds are deposited from solutions as a mixture of two types of crystals. The shapes of the crystals are related as mirror images. Pasteur separated by hand the two forms of sodium ammonium tartrate, a complex organic salt, and showed that solutions of the separated compounds rotated the plane of polarization of light in opposite directions. Any substance that rotates the plane of polarization of light is said to be optically active.

It would take us too far into physics to explain what is meant by the plane of polarization of light and how the rotation of the plane of polarization is measured. From our point of view it is sufficient to recognize that the detection of optical activity always indicates the presence of a molecule or structure that cannot be superimposed on its mirror image. The two mirror image forms of an optically active molecule are referred to as enantiomers. They are usually designated as D- and L-. Pasteur's experiment achieved the first separation of optical enantiomers, salts of D- and L-tartaric acid.

The simplest optically active molecules are those that include a carbon atom surrounded by four dissimilar groups at the apexes of a tetrahedron. The naturally occurring amino acids (except glycine) are of this type. Organic molecules in which each tetrahedral carbon atom is attached to at least two identical groups (e.g. hydrogen atoms) are optically inactive since they can always be

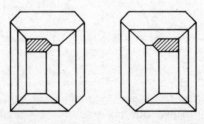

Figure 10A.2. D- and L-forms of sodium ammonium tartrate.

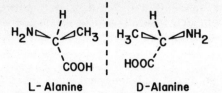

L- Alanine   D-Alanine

Figure 10A.3.   L- and D-alanine, showing that they are mirror images.

superimposed on their mirror images. Glycine, for example, is optically inactive, because its central carbon atom is attached to two hydrogen atoms.

Solutions of organic material can be optically inactive for one of two reasons. In some cases, all molecules in the solution may be inherently optically inactive, as in a solution of glycine. Alternatively, the solution may contain optically active molecules but the D- and L-enantiomers may be present in equal amounts so that the rotation produced by the D-enantiomer just cancels out that produced by the L-enantiomer.

All chemical reactions between optically inactive starting materials in solution give rise to optically inactive solutions of products. Sometimes, the product molecules are themselves individually inactive, but in the most interesting reactions optically active D- and L-molecules are produced but in equal numbers. This latter result is easily understood. In a symmetrical environment there is no possible reason why L-molecules should be produced more or less often than D-molecules.

Optically active crystals can sometimes be obtained from op-

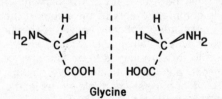

Glycine

Figure 10A.4.   Unlike alanine, glycine can be superimposed on its mirror image. If the right-hand molecule is rotated so that the NH₂ and COOH groups take up the same orientations that they have in the left-hand molecule, the two hydrogens of the right-hand molecule will coincide in position with those of the left-hand molecule.

161

Replicating Molecules and Natural Selection

tically inactive solutions, but this does not contradict the result stated above. In such cases, D- and L-crystals are always obtained in roughly equal numbers. This is what Pasteur found when he carried out his classic investigation of sodium ammonium tartrate crystals.

In a similar way molten inorganic substances that are themselves optically inactive sometimes solidify to form optically active crystals. Optically active crystals of the abundant mineral, quartz, are formed from an optically inactive melt of silica, for example. Whenever a careful count of left-handed and right-handed quartz crystals has been made, they have been shown to be present in almost equal numbers.

These results lead to a conclusion that is very important in discussion of the origins of life. While prebiotic synthesis may produce optically active molecules, they always produce the D- and L-enantiomers in equal number. No net optical activity could have been produced by abiotic reactions on the primitive earth. Some authors have correctly pointed out that this result is no longer strictly true when the magnetic field of the earth, the polarization of light from the sun, and other complications are taken into account. Most students of the field think that the importance of these factors has been exaggerated; we shall suppose that the prebiotic soup was optically inactive.

Living organisms are unique in that they contain a large number of optically active constituents. Only L-amino acids are present in proteins, for example, and only D-nucleotides in nucleic acids. Since the time of Pasteur, this remarkable feature of the composition of living things has fascinated biochemists and biologists. Pasteur himself wrote "I am inclined to think that life, as it appears to us, must be a product of the dissymmetry of the universe. . . ." Any complete theory of the origin of life must explain how organisms containing optically active biochemicals and polymers evolved from an optically inactive soup.

The solution to this problem is to be found by considering the interaction between pairs of optically active objects. We saw that although left-hand and right-hand gloves have the same shape, they interact differently with a right hand. However, each hand fits its own glove, and the two gloved hands that result are mirror images. Similarly, each hand fails to fit the wrong glove, and the resulting misfits form mirror images.

This example illustrates a general principle that is important in organic chemistry. When a pair of optically active molecules interact, they form one of a set of four possible combinations: if the two molecules are A and B, the possible products are D-A-D-B, L-A L-B,

162

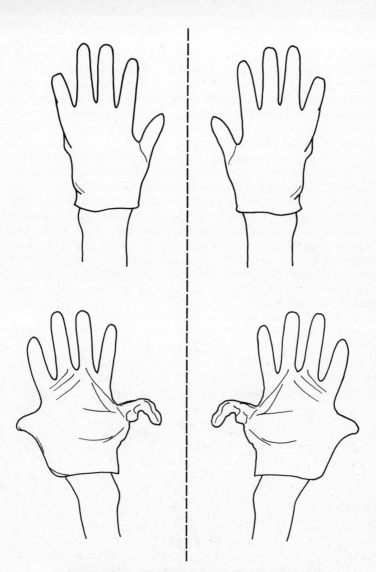

Figure 10A.5. Although a right-hand glove does not fit a left-hand glove and vice versa, the misfits obtained by putting a hand into the wrong glove are still mirror images.

D-A L-B, and L-A D-B. Just as with gloves and hands, D-A D-B and L-A L-B are mirror images and so are D-A L-B and L-A D-B, but L-A L-B and L-A D-B, for example, have quite different shapes. This argument can be extended to complicated systems containing many optically active components. If each component of a complex object is replaced by its mirror image, the mirror image of the object is obtained. If every molecule in the reader was replaced by its mirror image, it would be possible to construct the mirror image of the reader—with the heart on the wrong side, of course. However, if the configurations of some molecules were changed, but not of others, the result would be chaotic.

These arguments suggest that self-contained chemical systems that are perfect mirror images of each other behave identically, whereas systems in which some but not all components are mirror images have quite different chemical properties. This conclusion is correct and can be proved quite rigorously. It follows, for example, that if a naturally occurring enzyme (made up of L-amino acids) is able to synthesize a D-nucleotide, it is certain that the corresponding artifical enzyme made up of D-amino acids could synthesize the L-nucleotide. On the other hand, a corresponding polypeptide containing both D- and L-amino acids would almost certainly have no enzyme activity.

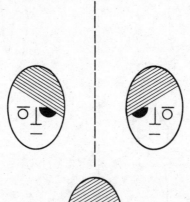

Figure 10A.6. Mirror images and partial mirror images.

We can now see that in trying to decide why living systems contain L-amino acids, D-nucleotides, and so on, we are trying to solve two quite different kinds of problems at the same time. To make progress, these problems will have to be separated. First, it will be necessary to explained why all the components of a protein or nucleic acid have the same configuration. More generally, the choice of the relative configurations of all the components of a living organism must be explained. Next one must elucidate why living organisms contain L-amino acids and D-nucleotides, rather than D-amino acids and L-nucleotides, that is, why a single apparently arbitrary choice was made between the living systems that we are familiar with and their mirror images.

From what is already known about the interactions of optically active molecules it can be seen that there could be no structural reason for selecting living organisms of one type of "handedness" rather than those of the other. It may, however, be anticipated that structural reasons will be found to account for most, if not all, of the choices of relative configuration that were made in the course of chemical evolution.

We do not yet understand what determined the choice of the relative configurations of all the different constituents of cells, but in some cases plausible explanations can be given. For example, it is quite easy to see why D- and L-nucleotides cannot occur together in the DNA structure. The regular double-helix composed of D-nucleotides spirals in a clockwise direction. If a DNA- like molecule were synthesized from L-nucleotides, it would differ from DNA only in that the structure would spiral in an anticlockwise direction. However, a molecule containing both D- and L-nucleotides could not form a regular structure at all, since the direction in which the helix turns would be constantly changing. An irregular structure would have properties quite different from those of DNA, so the mechanism that leads to the replication of DNA could not operate. A comparison with a spiral staircase may be useful here. Right-handed or left-handed spiral staircases are equally useful, but a staircase that was constantly changing its handedness would not be useful, except to a rock climber. (See Figure 10A.7).

At the moment there exists no convincing argument that explains why all amino acids in proteins have the same configuration, nor is it understood why L-amino acids, rather than D-amino acids, are associated with D-nucleotides. When more is known about the mechanism of protein synthesis, the answers to these questions should become clear. In the meantime, it may be supposed that, whenever, in the course of biochemical evolution, a new biochem-

Replicating Molecules and Natural Selection

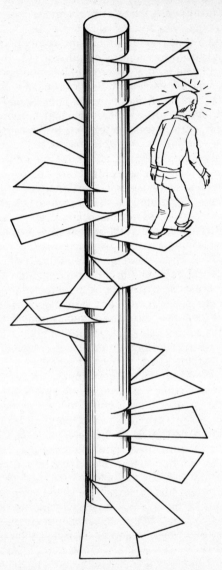

Figure 10A.7. The consequences of climbing a spiral staircase that changes from right-handed to left-handed direction of ascent.

ical compound became important, natural selection led to the choice of the enantiomer that functioned best.

To facilitate the next part of this discussion it is convenient to refer to contemporary living organisms as L-organisms (short for organisms with L-amino acids, D-nucleotides, and so on). We have to explain why the earth is populated entirely by L-organsims, while mirror image D-organisms are conspicuously absent. We have seen that there is no chemical reason why L-organisms should be more efficient than their mirror images, nor is there any obvious reason why D- and L-organisms should not coexist.

I believe that the success of L-organisms was a matter of chance; the earth could equally well have been populated by D-organisms. Some authors correctly claim that the influence of the earth's magnetic field, the rotational motions of the earth or some similar factor could have biased the odds in favor of L-organisms. It is doubtful that any of the biases so far suggested was significant.

Consider two models of the evolution of life on the primitive earth. First suppose that the evolution of a self-replicating system was a very improbable event. Then the first self-replicating system would have had ample time to eat up the prebiotic soup before any competitor could have gotten started. Once the prebiotic soup was used up, there was no longer any possibility of a second, independent origin of life. According to this model L-organisms inherited the earth because the first organism happened to be an L-organism. The first organism could equally well have been a D-organism; in that case we would now be wondering where the L-organisms had gone.

Next consider a completely different model in which we suppose that, once the prebiotic soup became sufficiently concentrated, self-replicating systems arose with high probability. Then, roughly equal numbers of L-and D-organisms would have evolved on the earth and would have competed against each other. This situation is very like the one that we consider in Chapter V when discussing the universality of the genetic code. L-organisms could only have eliminated their competitors by "discovering" one or more new adaptions that gave them a very pronounced selective advantage and allowed them to outgrow their mirror image competitors.

Perhaps we shall never know, as a matter of historical fact, just what happened three billion or more years ago to establish, once and for all, the predominance of L-organisms, but the problem does not raise any new matters of principle. The evolution of optical specificity was just part of the more general process of the establishment of biological order.

We saw in Chapter 5 that the universality of the genetic code forces us to ask how one type of biological organization came to win out over others that must have been very similar to it. The universality of L-organisms carried that question one step further—how did one form or organization win out over another that had exactly the same biological potential? Neither question can be answered in detail, but neither raised metaphysical issues.

From Replicating Polymers to Cells

11

The Evolution of Cells

Every living cell is surrounded by a selectively permeable membrane that functions, among other things, to hold together the components of the genetic apparatus, the biosynthetic enzymes and the small molecules that are utilized during growth and division. A simpler structure that played a role similar to that of the cell membrane must have evolved early in the development of life.

Before discussing the development of the cell, it is convenient to characterize a number of stages leading up to the evolution of self-contained biological systems. In the very first stage, individual polymer molecules must have formed and replicated at the expense of material that was already present in the prebiotic soup. Since there was, as yet, no requirement for cooperation between molecules, a segregating mechanism could have performed no useful function. Replication must have occurred in free solution, on the surface of rocks, or wherever else conditions happened to be favorable.

We believe that in the next stage of chemical evolution, nucleic acids began to direct the synthesis of other polymers, including polypeptides. As soon as this happened it became important to keep together whole families of macromolecules. At this point a structure was required that would adsorb or enclose macromolecules, but there was no need for a bag that was impermeable to small molecules. A membrane might, if anything, have been disadvantageous, since it could have excluded useful abiotic molecules from the interior of the "cell."

It was only when primitive organisms began making biochemical compounds for themselves that it became important to retain small molecules. The operation of natural selection leaves little room for charity. Most of the advantage that could be derived from the invention of new biochemical syntheses would have been lost if organisms had shared the benefits of their discoveries with their competitors. To prevent this from occurring it was necessary to invent a membrane that was impermeable to useful molecules made within the cell, but permeable to raw materials from outside. Such a membrane is said to be semipermeable.

All that is known about the replication of nucleic acids and the synthesis of proteins suggests that these processes evolved in a predominantly aqueous environment. When thinking about the origin of the cell, therefore, one needs to consider only those segregation mechanisms by means of which families of polymers are kept together in the presence of a large excess of water. We can propose several mechanisms of this kind that could have operated before the evolution of membranes, but it is not certain which of them was important on the primitive earth. Two of the most plausible mechanisms will be discussed: adsorption to mineral particles and adsorption to colloidal organic material.

Polymers stick very tightly to the surfaces of a number of common minerals. Proteins and nucleic acids are adsorbed strongly by the very common phosphate mineral, apatite; many organic substances, both small molecules and polymers, are adsorbed by various clays. It seems possible that the very first "organisms" consisted of nothing more than small mineral grains to which "genetic systems," that is families of replicating polymers, were adsorbed. "Reproduction" would have occurred whenever a family of polymers managed to colonize a new grain. If this picture is correct, the most primitive organisms were more like pin cushions than bags.

Theories of this kind have one great attraction. It has always been thought that the surface of a mineral might act as a primitive catalyst and help to bring about the orderly polymerization of amino acids and nucleotides. If the same mineral could both catalyze the formation of polymers and hold on to them once they had formed, it would have provided an advantageous site for further evolution. We know of minerals that catalyze polypeptide formation and of others that adsorb nucleic acids and proteins, but so far we do not know of any mineral that has all the properties that are required to favor extensive biological evolution. It is clearly important to search systematically for such a mineral.

An alternative and equally plausible theory suggests that cells evolved from some kind of colloidal particle, for example a coacervate of the type discussed briefly in Chapter 9. So little is known about the composition of the prebiotic soup that it is not possible to be precise about the chemical nature of this aggregate. Nonetheless, it seems likely that colloidal droplets of one kind or another would have formed in time from any sufficiently concentrated and sufficiently complex mixture of organic compounds. The cell, according to this theory, arose from a microscopic droplet of "oil" floating in the prebiotic soup.

A detailed proposal of this kind involves thermal polypeptides. If the polymers that are formed by heating a mixture of amino acids at 130°C (Chapter 9) are boiled with water, free-floating microspheres of organic material are formed. These microspheres are about the size of bacteria and they adsorb polynucleotides and many other organic molecules from aqueous solution. Dr. Sidney Fox has suggested that these microspheres are the precursors of living cells.

It is doubtful that thermal polypeptides formed the major component of the matrix from which cells evolved. However, they do provide an interesting illustration of the kind of structure that could have formed in the prebiotic soup. It seems likely that polypeptides and other polymers, no matter how they were formed, would have stuck together to give similar colloidal droplets. Once formed, these droplets could have collected a variety of polymers, metal ions, organic phosphates, and other molecules at their surfaces. Colloidal aggregates of this kind may indeed have played a part in the evolution of cells.

It is not known whether mineral particles or colloidal organic droplets were important for the first step in the evolution of the cell. Nevertheless, in either case, the next step must have been the evolution of a semipermeable membrane. Biological membranes are made up in large part of a class of organic molecules called lipids. Under some circumstances, artificial membranes are formed when lipids are dispersed in water (Chapter 4). If lipids had been present in the prebiotic soup, membranes could perhaps have formed spontaneously; otherwise it is much harder to envisage the steps that could have led from a colloidal aggregate to a cell surrounded by a selectively permeable membrane.

The Evolution of Metabolism

The earliest organisms grew in the prebiotic soup by making use of preformed molecules, such as nucleotides and amino acids. Many modern organisms, on the other hand, synthesize all biochemical compounds they need from a few very simple starting materials. The bacterium *E. coli,* for example, requires only glucose as a source of carbon, while plants thrive on carbon dioxide. This section will describe the stages by which primitive organisms became independent of the prebiotic soup.

Living organisms could not have derived any advantage by synthesizing for themselves compounds that were still freely available in the prebiotic soup. However, as soon as the supply of essential biochemical compounds was used up, alternative sources had to be found. It was at this point that organisms learned to synthesize these compounds from simple starting materials. The sequences of synthetic reactions that are used by living organisms to convert simple precursors to complex biochemical compounds will be referred to as biosynthetic pathways. Biosynthetic pathways differ from prebiotic pathways because they utilize enzymes. Consequently, every step in a biosynthetic pathway must proceed in solution.

Some biochemical pathways may have differed little from the corresponding prebiotic pathways; the first enzymes may have helped along processes that were occurring inefficiently in the absence of catalysts. However, this could not always have been the case, as the following example will show.

Phenylacetylene is readily formed from methane in the gas phase and reacts with aqueous ammonia and hydrogen cyanide, under appropriate conditions, to give the essential amino acid, phenylalanine. On the primitive earth it was probable that phenylacetylene formed in the atmosphere and reacted in oceans and lakes to form phenylalanine. However, phenylacetylene cannot easily be made in solution, even with the help of enzymes. When the supply of phenylalanine ran out, it was necessary, therefore, to invent a completely new synthesis of phenylalanine that did not involve phenylacetylene.

It is believed that many biosynthetic pathways developed, paradoxically, by reversing the pathways of spontaneous decomposition. This difficult idea is best illustrated by an analogy from modern industry. The Israeli air force was at one time largely dependent on France for its supply of fighter aircraft. After the war of 1967, France ceased to supply spare parts for these aircraft and Israel was forced to manufacture its own replacements. It has been claimed that it was the need to manufacture an increasingly wide range of spare parts that provided the incentive for the establishment of an autonomous aircraft industry in Israel.

This example illustrates a general principle: when a complex product of technology that has previously been supplied *from outside* is in short supply, it is usually easier

From Replicating Polymers to Cells

to improvise the repair of damaged models than to develop a complete production line. It is almost always possible to improvise repairs, whereas the manufacture of a product from raw materials is a major undertaking that requires extensive planning. Once enough repair capacity is available, it may become easier to develop a complete production line.

A similar idea is important in the context of biochemical evolution, because natural selection must achieve its results without planning. The development of a biosynthetic pathway that involves several steps is illustrated in Figure 11.1 When the supply of a compound B runs out, an ample supply of its first decomposition product D_1 will often be available. The first enzyme to evolve (E_1) will therefore convert D_1 to B. In time D_1 will be used up and it will be necessary to use D_2, the next decomposition product, as a starting point for the synthesis of B. However, since the enzyme E_1 is already available, it is only necessary to evolve a single enzyme E_2 to convert D_2 to D_1; then E_1 would finish the synthesis by converting D_1 to B. In a similar way new steps E_3, E_4, and E_5 could be added until it became possible to convert some very abundant precursor P through D_4, D_3, D_2, and D_1 to B. The evolution of the pathway would then be complete.

$$P \longleftarrow D_4 \longleftarrow D_3 \longleftarrow D_2 \longleftarrow D_1 \longleftarrow B \qquad \text{Spontaneous decomposition}$$
$$D_1 \xrightarrow{E_1} B \qquad \text{1st synthetic enzyme develops}$$
$$D_2 \xrightarrow{E_2} D_1 \xrightarrow{E_1} B \qquad \text{2nd synthetic enzyme develops}$$
$$D_3 \xrightarrow{E_3} D_2 \xrightarrow{E_2} D_1 \xrightarrow{E_1} B \qquad \text{3rd synthetic enzyme develops}$$
$$D_4 \xrightarrow{E_4} D_3 \xrightarrow{E_3} D_2 \xrightarrow{E_2} D_1 \xrightarrow{E_1} B \qquad \text{4th synthetic enzyme develops}$$
$$P \xrightarrow{E_5} D_4 \xrightarrow{E_4} D_3 \xrightarrow{E_3} D_2 \xrightarrow{E_2} D_1 \xrightarrow{E_1} B \qquad \text{5th synthetic enzyme develops}$$

$$P \xrightarrow{E_5} D_4 \qquad\qquad\qquad B \qquad \text{Useless}$$
$$P \xrightarrow{E_1 + E_2 + E_3 + E_4 + E_5} B \qquad \text{Impossible}$$

Figure 11.1. The top line shows the hypothetical course of decomposition of an important biochemical B. The next five lines show the "rescue" of successive decomposition products. The last two lines show that partial synthesis of D_4 from P is useless, while the direct route from P to B could never evolve.

It is important to notice that, in this scheme, each enzyme, as soon as it evolves, performs a useful function. There is no corresponding way of evolving a pathway from P to B in the forward direction, if the transformation cannot be achieved in a single step. An enzyme, for example E_4, could never be selected alone, because it would be useless without the other enzymes of the pathway. The chance of several enzymes evolving simultaneously is negligible.

The next phase of biochemical evolution must have occurred when even the most abundant components of the prebiotic soup were exhausted. It then became necessary to derive new, water-soluble organic materials from the gaseous components of the atmosphere; the photosynthetic fixation of carbon dioxide (Chapter 4) must have evolved at this time. Afterwards photosynthesis became essential for the continuation of life on earth. Organic compounds were constantly being oxidized to carbon dioxide and it was only through photosynthesis that the supply of organic carbon compounds could be replenished. It is likely that, at a still later stage, nitrogen-fixing organisms evolved, in response to a shortage of ammonia in the environment.

The evolution of cells that were surrounded by semipermeable membranes and were capable of carrying out a wide range of biosynthetic reactions marked the final stage in the evolution of life. The fossil evidence suggests that it occurred on the primitive earth at least three billion years ago. The evolution of multicellular organisms was still to take more than two billion years, while creatures with hard shells did not become abundant until half a billion years ago. Something is known and much more has been guessed about the history of life during the long period of specialization that marked the transition from primitive cells to complex modern organisms, but this material falls outside the scope of the present discussion. A number of the books recommended for further reading deal with the evolution of higher forms of life.

Natural Selection

12

Introduction

The evidence for evolution derived, for example, from the fossil record is now overwhelming. In the last decade or so, studies of protein sequences have further confirmed the traditional evolutionists' deductions in remarkable detail. Biochemists have shown that the difference between corresponding proteins of different species are proportional to the evolutionary distance between the species. The sequence of amino acids in the hemoglobin of the great apes is almost indistinguishable from that of human hemoglobin. On the other hand, there are substantial differences between human hemoglobin and that of less closely related mammals, such as mice or rabbits. The hemoglobin of birds is even more different from that of man.

It is possible to write computer programs that examine the sequences of proteins from different species and deduce the evolutionary relation between the species. No information is given to the computer except the amino acid sequences, but the computer prints out a family tree that matches closely the one deduced by comparative anatomists and palaeontologists. During the nineteenth century it was suggested that God had "planted" the fossil record to deceive wicked biologists; a supporter of that point of view would now have to suppose that God had also "fixed" the sequences of tens of thousands of proteins in hundreds of thousands of species, for no better reason.

The fact that evolution has occurred does not tell us anything directly about the mechanisms that have operated. The work of the pre-Darwinian evolutionists should not be underestimated. They correctly recognized evolution as an alternative to special creation. Their failure to arrive at the correct mechanism is a measure of the originality of Darwin and Wallace.

Lamarck believed that animals could pass on to their descendants the characteristics that they had acquired during their own lifetime. The ancestors of the giraffe, for example, benefited from the advantage of longer necks and somehow influenced the lengths of the necks of their children. This idea may not be attractive today but it was almost inevitable in the context of pre-Darwinian thought.

The world is full of animals that are obviously adapted to their own particular way of life. If no one created them, how could they be so well-adapted? Before Darwin, it was natural to think that someone or something had directed adaptation. If it wasn't an external God, something internal was the only alternative. This led some Lamarckians to concepts connected with the "striving" of species for improvement. Just as physicists invented a nonexistent ether because their common sense view of wave motion was inadequate, so biologists invented the inheritance of acquired characteristics because their common sense ideas on adaptation were inadequate.

It may be worthwhile to digress at this point to say something more about the inheritance of acquired characteristics. Although the theory had mystical implications there is nothing mystical about the theory itself—it just does not apply to the evolution of terrestrial plants or animals. It is possible to cut the tips off the tails of mice for a few genera-

Steps toward a Solution

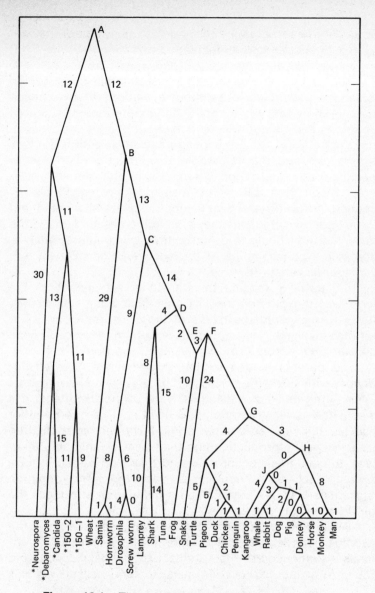

Figure 12.1. The relationship between species as determined from sequences of cytochrome C. Note how the computer program places species that are known on other grounds to be closely related close together in the family tree. Species marked with an asterisk are microorganisms. (Reproduced with permission from W. M. Fitch and E. Markowitz, *Biochem. Genetics,* **4,** 579, (1970).)

tions and then to examine the young to see whether they are born with shorter tails. In fact they still have normal tails, but if they did not, theoretical biologists would not be at a loss for nonmystical explanations.* (This example does not do justice to Lamarck's ideas about evolution—it was chosen only to show that the theory of the inheritance of acquired characteristics is not ridiculous.)

We have seen that the idea of natural selection is a very simple one and that it completely eliminates the need to postulate any internal or external "will" that directs evolution. Small heritable variations arise in individuals by chance; if they are disadvantageous they are eliminated, but if they are advantageous they enable the descendants of the fortunate individuals to outgrow their competitors. Adaptation is the consequence of the accumulation of many advantageous variations.

Let us now imagine the Darwinian explanation of the length of the giraffe's neck. Darwinians would claim that, had you measured the necks of the creatures from which giraffes evolved, you would have found that some were shorter than others. You would also have noted that long-necked parents tended to have long-necked children, *independently of their environment.* However, in environments where long-necked individuals were at an advantage, say where the lower vegetation had been eaten by some other species, tall well-fed parents would have had more children that reached reproductive age than shorter animals. Needless to say their children would have had longer-than-average necks—selection would have begun. The repetition of such selection, generation after generation, produced the giraffe's neck.†

According to Lamarck, evolution proceeds because the average neck length of the children of a given pair of animals is influenced by the parents' experience. According to Darwin, evolution proceeds because long-necked animals

---

* The invention of a molecular mechanism that permits the Lamarckian evolution of tail length in mice is left as an exercise for the advanced student.

† A more plausible but less graphic theory claims that the giraffe's legs and neck evolved to allow the animal to escape from lions.

have *more* of those children who, because of their longer necks, can compete efficiently for food and so survive to reproductive age. This is not a metaphysical distinction: in principle it would only be necessary to measure neck-lengths in a few families of evolving giraffes to show that Darwin's theory is correct. In practice this kind of experiment is rarely if ever possible, and we are forced to rely on less direct evidence.

Is There a Paradox about Evolution?

In the mid-1970s it is difficult to understand just how revolutionary the theory of evolution by natural selection seemed in the mid-nineteenth century. Today, to many biologists, the law of natural selection seems almost a tautology—at best it is a piece of theoretical biology that could be worked out by anyone able to multiply. One should not be deceived—it was a very hard discovery to make.

Darwin and Wallace are rightly regarded as among the greatest scientific innovators because they brought about a conceptual revolution. Darwin's solution to the central problem of biology was far from obvious even to his most brilliant supporters, until they had digested the accumulated evidence. T. H. Huxley was not completely convinced about natural selection until he read the *Origin of Species*. When he had finished he is reported to have exclaimed, "How exceedingly stupid not to have thought of that."

Darwin's work provided the solution to the central problem of biology—it showed how complex well-adapted organisms evolve from organisms that are simpler and less well-adapted. Recent experimental studies fully justify his belief that the variations that make selection possible are the consequences of random events. Why do so many philosophers claim that there is a paradox about evolution?

The supposed paradox is concerned with design. It is very hard to avoid using words that suggest purpose when describing the wonderfully adapted structures that occur in the living world. It is very tempting to describe the ribosome, for example, as an elaborate structure designed to carry out protein synthesis. Why are we tempted to use these "loaded" words when describing what, after all, are only the consequences of errors in the process of nucleic acid replication?

One view, the only correct one, points out that when biologists make statements that seem to involve design or purpose, they are using convenient abbreviations. When they say, "the ribosome was designed to carry out protein synthesis," they are avoiding long explanations. We would have to say, "early in the evolution of life, a nucleic acid that could direct the synthesis of informed polypeptides would have had a selective advantage over. . . ." Before we had finished, we would have rewritten much of Chapter 10. Perhaps we should agree to use the verb "to naturally select" in such situations. The phrase "the ribosome was naturally selected to carry out protein synthesis" may be inelegant, but it expresses what most biologists mean quite precisely.*

Although this is the formally correct answer, it will not satisfy everyone. It is genuinely surprising that an organism that has evolved by random mutation and selection appears to be designed. The idea is contrary to intuition. It is true, but to many it will always remain ridiculous.

Several of the greatest discoveries in science are counterintuitive. How many people *feel* that the earth is moving and the sun is at rest, or that objects continue in uniform motion until something happens to them? I suspect that most people have not assimilated the first idea intuitively. As you watch the sunset, do you *feel* the sun sinking in the West or yourself spinning away from it? The second idea feels more and more reasonable as the space program advances—you really do need a retro-rocket to bring the astronauts down. Evolution by natural selection is perhaps the hardest of the counterintuitive ideas that we are asked to accept. It is not only surprising, but, to some people, downright disagreeable.

Only experience leads to the assimilation of correct, counterintuitive arguments. One comes to accept them because one cannot do without them. If you read a lot about

* Darwin himself accepted this usage.

Steps toward a Solution

ribosomes or butterflies and think hard enough about the way they came to be as they are, you will probably find that you are using the idea of natural selection, without noticing it — alternatively, you may give up and become a mystic. It seems to me even more important to consider the origins of life in terms of natural selection; otherwise there is no way of approaching some of the most interesting aspects of the subject.

Once one gets used to the idea of natural selection, one finds it helpful in thinking about the development of many systems other than living organisms. One should not underestimate the importance of trial and error in the development of technology, for example. The development of the revolver is an object lesson in evolution. Those clumsy guns with revolving barrels are the dinosaurs; there were many small successes and many great failures on the way to the Peacemaker, and for every company that presently makes revolvers, there must be ten that have been eliminated.

There is no paradox about evolution by natural selection. The idea is a subtle one and it takes a little time to get used to it. No doubt that there is much more to be learned about the mechanism of evolution, but most biologists would agree that two things are certain — biological adaptation is made possible by natural selection, and all evolution depends on random events. The replacement of "will" by "chance" as the mediator of biological change has transformed our view of man's relation to the rest of the Universe. For better or worse, that transformation is unlikely to be reversed.

In a precisely parallel way we know that only processes involving natural selection could have led to the development of life on earth. We have even more to learn about the origin of life than about the origin of species. We cannot even be quite sure that molecules like nucleic acids were the first materials on which selection acted. Nonetheless, I think we are justified in rejecting any theory of the origins of life that fails to explain at least in principle, how natural selection could have led to the development of biological complexity.

PART
THREE

EXTRATERRESTRIAL LIFE

What is Life?

13

Introduction

In Chapters 15 and 16 a set of related problems concerned with the existence of life on other planets will be discussed. Does life exist elsewhere in the solar system? Could life on another planet be based on silicon compounds rather than on carbon? Are there intelligent societies elsewhere in the universe? Problems of this kind inevitably raise questions about the nature of life. In this chapter this age-old question will be taken up.

The word "life" has many different but overlapping meanings that have developed independently of systematic biology. Hence, it is not well-adapted to use in technical, scientific discussions. Unfortunately, there are no precisely defined English expressions to substitute for it; we cannot discuss life on other worlds without saying something about "life."

The following examples illustrate the kinds of confusion that can arise. As far as the layman is concerned, a horse is either dead or alive, and that is the end of the matter. The biologist's point of view is more complex. Of course he would not deny the fundamental difference between a dead horse and a live horse but he would regard the former as still part of the living world, as long as most of its cells were functioning normally. In a quite different context, many elementary introductions to biology state that all living things must be able to reproduce, but, if interpreted literally, this requirement would "kill" the mules, since mules are sterile. Even biologists themselves are liable to argue about viruses, since viruses can reproduce, but only with the help of a living host. We shall see that these difficulties arise because many of the questions that are asked about the nature of life are not scientific, but are concerned with English usage.

Today most scientists would agree that "What is life?" is not a good scientific question. Scientists like to deal with questions that can be answered by experiment in an impersonal way. "What is life?" is not a question of this kind, for it has something in common with questions like "What is freedom?" It is quite possible for two microbiologists to reach identical conclusions about the structure and function of a virus, for example, and yet to disagree as to whether it is living or not. Similar difficulties are likely to arise when we try to decide whether a society is free or not.

To help the reader to distinguish more clearly between the scientific and the semantic issues involved, consider the following hypothetical situation. Suppose we came upon a strange "organized body" on Mars and had to give an account of it to our scientific colleagues. We would first examine the structure of the object and how it behaved. Once we could describe the object and its behavior in complete detail, we would have finished our work as "biologists." If the object were a very strange one, we might find ourselves trying to decide whether or not to call it living. In that case, we would not be carrying on a normal scientific enquiry, but trying to settle the way in which the English words "living" and "nonliving" should be used in a novel context.

This suggests that when we discuss the forms that life might take on other planets, we should concentrate on the structure and behavior of strange objects rather than their status as "living" or "nonliving" beings. Let us start by examining the attributes of terrestrial living organisms. This will lead us to characterize a broad class of objects that includes all those that are considered alive on intuitive grounds. However, this class also includes a number of simpler, self-reproducing objects that would puzzle the layman because he would not know whether to call them living or not. For reasons that will become clear all such objects will be referred to as CITROENS.

Terrestrial Biology

Most elementary introductions to biology contain a section on the nature of life. It is usual in such discussions to list a number of properties that distinguish living from nonliving things. Reproduction and metabolism, for example, appear in all of the lists; the ability to respond to the environment is another old favorite. This approach extends somewhat the chef's definition "If it quivers, it's alive." Of course, there are also many characteristics that are restricted to the living world but are not common to all forms of life. Plants cannot pursue their food; animals do not carry out photosynthesis; lowly organisms do not behave intelligently.

It is possible to make a more fundamental distinction between living and nonliving things by examining their *molecular* structure and *molecular* behavior. In brief, living organisms are distinguished by their *specified* complexity.* Crystals are usually taken as the prototypes of simple, well-specified structures, because they consist of a very large number of identical molecules packed together in a uniform way. Lumps of granite or random mixtures of polymers are examples of structures which are complex but not specified. The crystals fail to qualify as living because they lack complexity; the mixtures of polymers fail to qualify because they lack specificity.

* It is impossible to find a simple catch phrase to capture this complex idea. "Specified and, therefore, repetitive complexity" gets a little closer (see later).

These vague ideas can be made more precise by introducing the idea of information. Roughly speaking, the information content of a structure is the minimum number of instructions needed to specify the structure. One can see intuitively that many instructions are needed to specify a complex structure. On the other hand, a simple repeating structure can be specified in rather few instructions. Complex but random structures, by definition, need hardly be specified at all.

These differences are made clear by the following example. Suppose a chemist agreed to synthesize anything that could describe accurately to him. How many instructions would he need to make a crystal, a mixture of random DNA-like polymers or the DNA of the bacterium *E. coli?*

To describe the crystal we had in mind, we would need to specify which substance we wanted and the way in which the molecules were to be packed together in the crystal. The first requirement could be conveyed in a short sentence. The second would be almost as brief, because we could describe how we wanted the first few molecules packed together, and then say "and keep on doing the same." Structural information has to be given only once because the crystal is regular.

It would be almost as easy to tell the chemist how to make a mixture of random DNA-like polymers. We would first specify the proportion of each of the four nucleotides in the mixture. Then, we would say, "Mix the nucleotides in the required proportions, choose nucleotide molecules at random from the mixture, and join them together in the order you find them." In this way the chemist would be sure to make polymers with the specified composition, but the sequences would be random.

It is quite impossible to produce a corresponding simple set of instructions that would enable the chemist to synthesize the DNA of *E. coli.* In this case, the sequence matters; only by specifying the sequence letter-by-letter (about 4,000,000 instructions) could we tell the chemist what we wanted him to make. The synthetic chemist would need a book of instructions rather than a few short sentences.

It is important to notice that each polymer molecule in a random mixture has a sequence just as definite as that of *E.*

Extraterrestrial Life

coli DNA. However, in a random mixture the sequences are not specified, whereas in *E. coli,* the DNA sequence is crucial. Two random mixtures contain quite different polymer sequences, but the DNA sequences in two *E. coli* cells are identical because they are specified. The polymer sequences are complex but random; although *E. coli* DNA is also complex, it is specified in a unique way.

The structure of DNA has been emphasized here, but similar arguments would apply to other polymeric materials. The protein molecules in a cell are not a random mixture of polypeptides; all of the many hemoglobin molecules in the oxygen-carrying blood cells, for example, have the same sequence. By contrast, the chance of getting even two identical sequences 100 amino acids long in a sample of random polypeptides is negligible. Again, sequence information can serve to distinguish the contents of living cells from random mixtures of organic polymers.

When we come to consider the most important functions of living matter, we again find that they are most easily differentiated from inorganic processes at the molecular level. Cell division, as seen under the microscope, does not appear very different from a number of processes that are known to occur in colloidal solutions. However, at the molecular level the differences are unmistakable: cell division is preceded by the replication of the cellular DNA. It is this genetic copying process that distinguishes most clearly between the molecular behavior of living organisms and that of nonliving systems. In biological processes the number of information-rich polymers is increased during growth; when colloidal droplets "divide" they just break up into smaller droplets.

CITROENS

We have seen that it is difficult, if not impossible, to find a rigid definition of life that incorporates all of our intuitive ideas. Instead, we can make a list of the attributes that help us to decide whether or not a system is living—reproduction, metabolism, excitability, and so on—and agree to call an organism alive if it possesses a suitable selection of these attributes. This is a useful approach in introductory discussions of terrestrial biology, but it is not so useful when one discusses alien forms of life. In the latter case, it is too difficult to complete the list; it is impossible to enumerate all the types of behavior that might characterize nonterrestrial forms of life.

The discussion in the previous section suggests a more useful procedure. When we talk about life on other planets, we are not thinking about simple objects that behave in a simple way. The structure and behavior of an object would need to be nonrandom and reasonably complicated to interest the student of extraterrestrial life. We have already seen that a great deal of information is needed to specify the structure of a complicated nonrandom object. It may be concluded that anything that we would want to call "living" would have a high information content. This apparently simple requirement has far-reaching consequences.

It follows immediately that any "living" system must come into existence either as a consequence of a long evolutionary process or a miracle. It can be shown that the probability of a structure arising spontaneously decreases very rapidly as the information content of the structure increases. A chimpanzee sitting at a typewriter might easily compose a word of Shakespeare, but would be unlikely to compose a line; if a chimpanzee wrote a whole play, the event could legitimately be called a miracle. In a similar way, the formation of a structure complex enough to be called "living", in a single event, would be a miracle.

Since, as scientists, we must not postulate miracles we must suppose that the appearance of "life" is necessarily preceded by a period of evolution. At first, replicating structures are formed that have low but non-zero information content. Natural selection leads to the development of a series of structures of increasing complexity and information content, until one is formed which we are prepared to call "living".

This conclusion, in turn, has important consequences. Since selection cannot occur without reproduction, no system of "living" organisms can evolve except from primitive replicating structures.* This does not mean that every advanced organism must be able to reproduce, but the persistence of "life" requires that some of them can. Every "living" creature must be the descendant of other related creatures.

Finally, during reproduction, the information that specifies the structure of the parent must be passed on to each child; otherwise the children would not resemble the parents. The transmission of this genetic information must be reasonably accurate, for if too many errors occurred the children would not be able to survive. It follows that genetic information must be stored in a structure that is stable throughout the reproductive lifetime of the parent.

These considerations lead us to the following requirements that are necessary and sufficient to qualify a structure as "alive":

1. The object is complex and yet well-specified.
2. The object is able to reproduce.†

These conditions, as we have seen, imply that:

(a) the object is a product of natural selection,‡
(b) the information needed to specify the object is stored in a structure that is stable for the reproductive lifetime of the object.

A new term for such "living" organisms, whether terrestrial or not, must now be introduced. They are Complex Information-Transforming Reproducing Objects that Evolve by Natural Selection—CITROENS. (NOTE this noun has no singular form—one cannot have a citroen, only a family of CITROENS).

Any objects that we would intuitively call alive would

* Such structures are not necessarily based on stable molecules, although these structures are easier to think about than any others (see below). The stable structures could conceivably consist of patterns of chemical reactions, for example.
† Alternatively, the object may be the descendant of related objects that can reproduce, even if it is itself "sterile."
‡ Unless it is a product of human technology—see p. 196.

have to be CITROENS. However, DNA-like molecules that were capable of replicating, say on the surface of a catalytic mineral, would also be examples of CITROENS. All CITROENS are interesting because they can evolve by natural selection; even if they are themselves relatively simple they are able, under suitable conditions, to generate systems of great complexity. Any intelligent form of life on any planet must have evolved in the past from a system of simple CITROENS.

Novel CITROENS

These ideas lead us naturally to a useful classification of possible types of biological organization. If the information content of a family of CITROENS is stable, it must be stored in some stable structure. We are familiar with only one type of stable "genetic memory", the nucleic acid system. The literature of science fiction abounds with alternative "genetic-memory banks". A number of possibilities are suggested by computer technology; a genetic memory might be based, for example, on a system of electric currents in a superconductor, or on the pattern of magnetization of a magnetic tape. Another favorite theme is that of submicroscopic "nuclear life", based on some stable reproducing collection of protons, mesons, neutrons, and other fundamental particles. Hoyle has described a living organism that has a continuous cloud-like structure.

The reader should appreciate that these ideas, and many others like them, remain in the field of science fiction. It is not clear that there are any stable, self-reproducing "genetic-memory banks," except those based on stable chemical structures like nucleic acids. On the other hand, we cannot prove that they are impossible. Maybe, on some other planet, the local scientists, while conceding the possibility of genetic systems based on molecules like nucleic acids, doubt that they would really work.

It will be left as an exercise for the reader to work out the details of a few of the more bizarre systems. Alternatively, he might like to read the science fiction referred to in the bibliography. Let us now direct our ideas toward earth-like CITROENS, by restricting further discussion to "genetic-memory banks" based on stable chemical structures.

For a polymeric molecule to act as a genetic material, it must be stable throughout the life of the organism and must also be able to direct the synthesis of identical or nearly identical genetic molecules from substances in its environment. We are familiar with a single example—the nucleic acids—but quite different systems may be possible.* I can see no reason why genetic systems based on materials that are stable at higher temperatures, perhaps silicates, should not function on planets as hot as Mercury. Similarly, materials that would react explosively under terrestrial conditions might form the basis for a stable genetic system operating at very low temperatures. We are not yet in a position to say whether high-temperature CITROENS could be constructed from derivatives of silicon or boron, for example, or if low-temperature CITROENS could evolve and thrive at very low temperatures in an ocean of liquid hydrogen.

It is only when we come to CITROENS that are based on carbon chemistry and that survive in an aqueous environment, that we are on more familiar ground. Even then, there is no reason to believe that an independent form of life of this kind would resemble terrestrial life very closely. It is not clear that all genetic systems need be based on polymers containing phosphorus, for example, or that all genetic molecules need to contain nitrogen in addition to carbon, hydrogen, and oxygen. Perhaps, a single family of macro-molecules might act both as the genetic structure and the enzymatic machinery of nonterrestrial CITROENS.

We do, however, have experimental evidence that tells us something about the probable composition of earth-like CITROENS. We know that the range of organic compounds formed abundantly under prebiotic conditions is quite narrow. It seems probable that earth-like life, if it exists elsewhere, will likely be based on the kinds of compound that are formed in reducing planetary atmospheres. In the light of the evidence presented in Chapters 6–8, it would be sur-

* One such system is discussed in detail in the book by Cairns-Smith cited in the bibliography.

prising, therefore, if amino acids and sugars failed to play some role in the chemistry of other earth-like forms of life.

It would be rash to draw more far-reaching conclusions. It is always tempting to believe that the details of terrestrial biochemistry are the inevitable consequences of the laws of chemistry and physics and depend little on historical accident. This temptation should be resisted. We can safely predict that if we ever encounter extraterrestrial CITROENS, they will have many unearthly features.

The Products of Human Technology

William Paley (1743–1805) in his *Natural Theology, or the Evidence of the Existence and Attributes of the Deity from the Appearance of Nature* drew attention to a subtle and interesting problem. Paley first remarks that if we discovered a watch lying on the ground, we would not question that it had a maker. We could not believe that a complicated object so obviously adapted to telling the time is the product of the blind forces of inorganic nature. Paley then argues, by analogy, that the living world must also have a maker, God, since it shows even greater evidence of design.

We have already seen that the operation of natural selection, a completely random process, leads to the evolution of organisms that do indeed seem to have been designed. Paley made the mistake of thinking that there is a single way in which complex, well-adapted objects came into existence—creation. In fact there are two that we now know about—natural selection, and fabrication by man. However, Paley was right to emphasize the need for special explanations of the existence of objects with high information content, for they cannot be formed in nonevolutionary, inorganic processes.

We are familiar with many products of technology that fail to be CITROENS only because they do not reproduce autonomously. There is, thus, an exception to the rule that objects of high information content must be the direct product of natural selection; they may be the products of human ingenuity. This exception does not weaken the argument, since the intelligent "creators" in this case are themselves the products of natural selection.

The creation of life in the laboratory is a subject that fascinates the layman. If we are prepared to count the simplest CITROENS as alive, I think that the creation of totally new forms of life will be possible in the future, perhaps within a hundred years. Of course, any man-made CITROENS will be far removed from the little green men of the more simple-minded science fiction.

Extraterrestrial Organic Chemistry

14

The Murchison Meteorite

Many solid objects fall from space into the earth's atmosphere each year. The heat generated as they are slowed down by the frictional resistance of the air is sufficient to make them white hot. The smaller objects are vaporized completely in the atmosphere, but some of the largest ones reach the surface of the earth. Incandescent objects that vaporize completely in the atmosphere are called meteors (shooting stars); objects that reach the surface of the earth are called meteorites.

Large meteorites are quite uncommon; it is estimated that objects with diameters much greater than a meter (a yard) reach the surface of the earth only about once in ten years. A number of large craters formed by the impact of meteorites are known; they show that a few massive objects must have struck the earth long ago. In more recent times a meteorite that fell in Siberia caused an explosion powerful enough to be detected all over the earth.

We do not know, for sure, where meteorites come from, nor even if they all come from the same place. It is most likely that they originate in the asteroid belt, a collection of massive rocks and smaller debris that orbits around the sun between Mars and Jupiter. The material that makes up the asteroid belt is sometimes assumed to be the remains of a small planet that disintegrated long ago, but it is by no means certain that the asteroids were formed in this way.

The meteorites that are commonly displayed in muse-

Figure 14.1. Barringer's Crater in Arizona. Nearly a mile in diameter it was produced about 20,000 years ago by an iron-nickel meteorite weighing many thousands of tons. (Reproduced with permission from *Physics of the Earth* by T. F. Gaskell, Funk and Wagnalls Publishing Co., New York, 1970.)

ums are made of metallic iron and have a very characteristic appearance. The stony meteorites resemble ordinary terrestrial rocks. They are, therefore, much harder to find and they make much less impressive exhibits. Although stony meteorites are not familiar to the public, they are, in fact, the most common variety.

In this chapter we shall be concerned with the carbonaceous chondrites, a subclass of the stony meteorites. The chondrites are named after the small spheres (chondrules) of silicate materials that they contain. The carbonaceous chondrites are uniquely interesting because up to 4% of their mass is carbon, some of it in the form of organic compounds.

The carbonaceous chondrites conduct heat poorly. Thus, although their surfaces are heated briefly to incandescence, their interiors remain cool as they fall through the atmosphere. In principle, therefore, it should be possible to identify the organic compounds native to a carbonaceous chondrite by removing the thin fusion crust and analyzing the unmodified material in the interior.

The Swedish chemist Berzelius attempted precisely such an analysis as early as 1834. At that time the techniques of organic chemistry were not sufficiently developed to permit Berzelius to identify particular organic compounds. He did, however, conclude that nonterrestrial organic material was present. Berzelius was careful to point out that the presence of organic material should not be interpreted to mean that living organisms necessarily existed at the site of origin of the meteorite.

Since Berzelius' time, organic chemists have tried to improve on his analysis. Although many organic compounds have been identified in one meteorite or another, none of the earlier analytical results can be accepted without reservation. The major problem has always been terrestrial contamination. Carbonaceous chondrites, because they are permeable to water, readily pick up organic compounds such as amino acids from the soil. Some of the most famous meteorites have been kept in museums and laboratories for years, and have picked up further contaminants during storage.

From time to time, a good deal of excitement is gen-

Extraterrestrial Organic Chemistry

erated by claims that the remains of living organisms have been identified in a meteorite. Sometimes, the "extraterrestrial organisms" turn out to be spores picked up on the ground; in other cases the organisms are probably inorganic artifacts. In no case is there evidence adequate to establish the presence of living organisms or their remains in a meteorite.

Although much of the organic material reported in earlier analyses of meteorites was picked up on the earth, the carbonaceous chondrites undoubtedly contain indigenous organic compounds. No doubt, some of these compounds have been identified correctly. However, until uncontaminated meteoritic material became available, it was very difficult to be sure whether an organic compound in a meteorite was extraterrestrial or not.

On September 28th, 1969, a large carbonaceous chondrite fell and broke up in an arid region close to the town of Murchison in Australia. Several pieces of the meteorite were collected within a few days, and many more within six months. The danger of contamination was fully appreciated, and every precaution was taken to avoid contact between the specimens and sources of terrestrial organic matter. By 1970 an extensive analysis of fragments of the Murchison meteorite was under way in two independent laboratories in the U.S.A.

Both laboratories confirmed that the Murchison meteorite contains, in addition to other organic compounds, large quantities of amino acids. The natural amino acids, glycine, alanine, glutamic acid, valine, and proline were identified. In addition, some simple amino acids that do not occur in the proteins of living organisms were found in the meteorite.

We have seen that in this instance the circumstances surrounding the collection of the specimens were such that the danger of contamination was minimized. The following three independent lines of evidence tend to confirm that the amino acids are extraterrestrial in origin.

First, the amino acids isolated from the meteorite include several that are absent from living organisms and others that are uncommon. These amino acids are the ones that, on theoretical grounds, would be expected in a mixture

Extraterrestrial Life

of prebiotic chemicals. In one or two cases, they had been identified among the products of prebiotic reactions before they were found in the Murchison meteorite. In more recent experiments all of the amino acids identified in the meteorite have been obtained by the action of an electric discharge on a "prebiotic" reducing atmosphere.

The second observation is the most convincing. We saw in the Appendix to Chapter 10 that the amino acids present in living organisms are optically active; they are almost always in the L-form. Amino acids in the soil are always of biological origin and, therefore, are optically active. On the other hand, prebiotic amino acids are optically inactive; D- and L-molecules are present in equal numbers. The amino acids in the Murchison meteorite are all optically inactive and hence cannot be terrestrial, biological contaminants picked up from the soil.

The final observation is concerned with isotope ratios. In brief, all terrestrial organic material of biological origin is characterized by a particular value of the ratio of the abundances of two carbon isotopes, ^{12}C and ^{14}C. The material on the meteorite is claimed to have a quite different ratio of ^{12}C and ^{14}C abundances. If the $^{12}C:^{14}C$ isotope ratio is indeed anomalous for the amino acids in the Murchison meteorite, they must be extraterrestrial in origin.

The Murchison meteorite contains a number of other organic compounds in addition to the amino acids, but they are not closely related to important biochemical compounds. The hydrocarbons found in the meteorite are of the kind that one would expect to form under prebiotic conditions, and they are quite different from those found in contemporary organisms. It is interesting also that important classes of biochemical compounds are absent from the meteorite; a careful search failed to detect adenine, for example. Although pyrimidines were found, they are not the ones that occur in nucleic acids.

The discovery of large amounts of several naturally occurring amino acids in the Murchison meteorite clearly establishes that these compounds are formed spontaneously somewhere else in the solar system. However, until we know more about the origin of the meteorite, it is not possible to draw any more precise conclusions. There is no guarantee that the amino acids found in the meteorite were formed under any of the prebiotic conditions that have been used in the laboratory. We cannot, therefore, claim that the new results support any detailed theory of prebiotic synthesis. On the other hand, they do encourage the belief that our general ideas about prebiotic synthesis have some validity.

Molecules in Space

The detection of organic molecules in regions of space very far from the earth has been made possible by recent developments in radioastronomy. Many molecules emit (or absorb) radio waves at characteristic sets of frequencies. Each molecule has, so to speak, a unique radiofrequency "signature" or spectrum that distinguishes it from all other molecules. Thus, if we can show that all of the frequencies in the "signature" of a molecule are among the frequencies emitted by a distant radio source, we can be sure that the molecule is present in the source. The improved radiotelescopes that have become available during the last decade or so have made it possible to apply radiofrequency spectroscopy to some distant astronomical objects.

Using this technique, many small molecules have been shown to be present in large amounts in interstellar dust clouds. The first molecules to be detected were water and ammonia, two molecules that were expected to be abundant. The next compounds to be identified came as a surprise to the astronomers. They were hydrogen cyanide, formaldehyde, and cyanoacetylene, a group of reactive organic molecules. Since then a number of other simple organic substances have been detected.

We are already familiar with hydrogen cyanide, formaldehyde, and cyanoacetylene; they are believed to have been amongst the most important prebiotic intermediates that were formed in the primitive reducing atmosphere of the earth. It is known that hydrogen cyanide, formaldehyde, and cyanoacetylene react in aqueous solution to give a variety of important biochemical compounds; sugars can be derived from formaldehyde, amino acids and adenine from hydrogen cyanide, and pyrimidine bases from cyanoacetylene. The presence of hydrogen cyanide and formaldehyde in the dust clouds is perhaps not too surprising, since they are among the simplest of organic molecules. However, cyanoacetylene is a rather unfamiliar substance, and its presence in the dust cloud was quite unexpected.

It is probably a coincidence that the first three organic molecules to be detected in space are important prebiotic intermediates. A number of further molecules have now been identified. Not all of these substances are known to be involved in the prebiotic synthesis of important biochemical compounds. Nevertheless, these novel discoveries in radio-astronomy provide another hint that rather few classes of organic compounds are formed under prebiotic conditions.

The dust clouds in which organic molecules have been detected are believed to be the sites of formation of new stars. Could similar compounds in the dust cloud from which the solar system evolved have contributed to the origins of life on earth? It was argued in Chapter 6 that preformed organic compounds in the dust cloud are unlikely to have been very important for the origins of life, because they would have been distributed uniformly throughout the earth.

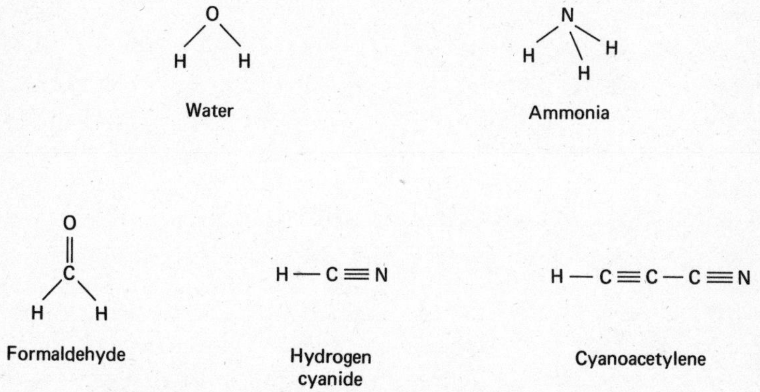

Figure 14.2. The first five molecules (other than H_2 and CO) discovered by radio-astronomers in space. Also five of the most important prebiotic molecules.

However, the organic matter that formed in the dust cloud and accumulated at the surface of the earth, may have played some part in the origins of life.

The new results described in this chapter tell us little about the details of prebiotic synthesis on the primitive earth, but they are very important in a wider context. We now know that molecules that can be synthesized under prebiotic condition in the laboratory and are thought to have been important for the origins of life on earth occur in two quite different extraterrestrial materials. This suggests that they are likely to occur at many places in the universe. Thus, when we come to consider the possibility that life has evolved on other planets, we shall be able to assume with some confidence that amino acids and related biochemical compounds were formed elsewhere in the universe.

Life in the Solar System

15

Historical Introduction

The possibility that there are living beings elsewhere in the universe has fascinated men for at least two thousand years. The Greek writer Lucien, who lived in the second century A.D., sent his adventurers on voyages to the moon and stars. He was not interested in problems of propulsion; the wing of an eagle and the wing of a vulture were all that his heroes needed to make their journeys. Lucien founded a literary tradition in which the behavior of the fictional inhabitants of the moon and planets is used to parody human behavior.

For Giordano Bruno, one of the most important scientists of the sixteenth century, extraterrestrial life was not a subject for light-hearted treatment. He wrote, "There are innumerable suns and innumerable earths circling round their suns just as our seven planets circle round our sun. Living things dwell on these worlds." Such speculations still sound daring—in Bruno's day they helped lead to the stake. He was burned by the Inquisition in the year 1600.

The first modern science fiction tale was written not long afterwards by Kepler, probably a little before 1635. Its publication was not thought prudent until after Kepler's death. Kepler is, of course, famous for having discovered the laws describing the motion of the planets around the sun. His *Somnium* or *Dream,* a fantasy written late in life, is less well-known, although it is a highly original work. It introduces for the first time many of the elements that characterize the best of modern science fiction.

The moon which Kepler describes is the moon which he could see through his telescope. Kepler maintains the plausibility of his narrative by the authenticity with which he describes the journey to the moon and the conditions that his hero, Durocatas, finds there. Kepler realized that the inhabitants of the moon must have adapted to the extremes of the lunar climate; his gigantic lizards grow to maturity and die in a single lunar day.

After the publication of Kepler's *Somnium,* the "plurality of worlds" became a popular topic of discussion in Europe, and has remained so to the present day. John Wilkins (1614–1672), one of the founders of the Royal Society of London, examined the possibility that there are intelligent beings on the moon and predicted that men would one day travel there. Christian Huygens (1629–1695), the Dutch physicist, considered each of the planets in turn as a possible habitation for life. Athanasius Kircher (1601–1680) concerned himself with quite different questions. He wondered, for example, whether water on the planets should be used for carrying out baptisms.

The first claim to have detected an extraterrestrial culture appeared at the end of the nineteenth century. Percival Lowell, an American astronomer, identified a series of linear features on the surface of Mars as canals built by the Martians. His work excited a great deal of public interest. In time it became clear that much of the Martian surface structure that Lowell described is imaginary, and that the rest can be explained in a quite ordinary way.

The last twenty years have seen a great expansion of

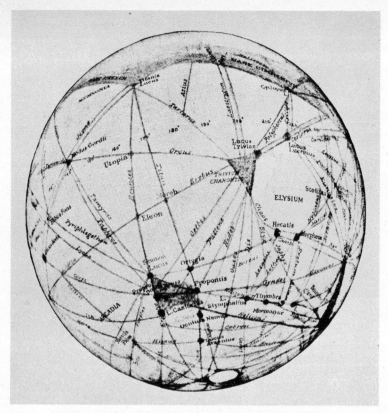

Figure 15.1. The canals of Mars according to Percival Lowell. No such features are revealed by more recent examinations either from the Earth or from orbiting space-craft. (From *Mars as the Abode of Life* by P. Lowell, Macmillan, New York, 1909.)

scientific interest in the possibility of extraterrestrial life. The biggest single factor has been the space program. The Viking mission to Mars in 1975, for example, is specifically designed to search for evidence of life. Since it represents an investment of three-quarters of a billion dollars, it is clear that many people take the possibility of life on Mars seriously.

The development of radioastronomy has also been im-

Life in the Solar System

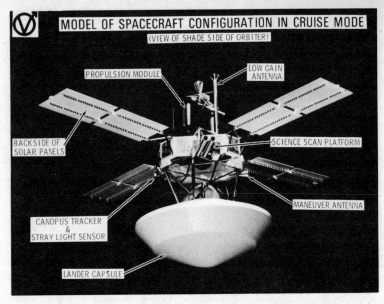

PROPULSION MODULE

LOW GAIN ANTENNA

BACKSIDE OF SOLAR PANELS

SCIENCE SCAN PLATFORM

MANEUVER ANTENNA

CANOPUS TRACKER & STRAY LIGHT SENSOR

LANDER CAPSULE

Figure 15.2. A Model of the Viking Spacecraft that will land on Mars in 1976. (Reproduced with permission of the National Aeronautics and Space Administration.)

portant. It is now possible to search for radiocommunications from planets in other solar systems. Astronomers, at least, seem to think that there is a chance that an extensive search would be rewarded; more than half of those who answered a recent questionnaire thought that it would be sensible to use large radiotelescopes to search for technologically advanced extraterrestrial cultures.

Life on the Moon and Planets

Our nearest neighbor is, of course, the moon. The moon is too light to retain an atmosphere, and there can be no liquid water on its surface. For these reasons almost everyone agreed that life was not likely to exist there. However, this did not prevent the development of a very strange controversy. Most biologists were prepared to ignore the possibility that organisms could survive on the moon. A small minority felt that although it was unlikely that there was life on the moon, lunar organisms, if they existed, might do so much damage to the population of the earth that extreme precautions should still be taken to prevent returning Apollo astronauts from infecting the earth.

Faced with this difficult choice, the National Aeronautics and Space Administration (NASA) decided on caution. An elaborate and expensive space quarantine was built in Texas, and returning astronauts and their equipment were isolated until shown to be free of dangerous lunar organisms. We now know that the surface of the moon is sterile. The regions examined so far are not only lifeless, they are also almost free of organic carbon. Almost everyone agrees that there is no chance that life exists anywhere on the surface of the Moon.

With the advantage of hindsight, we can now see that the lunar quarantine was unnecessary. At the time it was very difficult to know what to do about a contingency that was extremely improbable, but might just possibly have been totally disastrous for the human race.

It is generally believed that the evolution of earth-like life would not be possible at high temperatures in the absence of liquid water. Hence it is thought extremely unlikely that the planet Mercury is inhabited since Mercury is very hot and has no atmosphere. It is harder to be certain about the lower limit of the temperature range in which life is possible, but it is likely almost certain that the outer planets, Uranus, Neptune, and Pluto are too cold to support life.

The surface of Venus is very hot ($\sim 480°C$). No liquid water could exist at this temperature; in fact few if any organic compounds could survive, even in the solid state. We can, therefore, be confident that there is no earth-like life at the surface of Venus. Since the top layer of the atmosphere is at a much lower temperature ($-40°$), there is certainly a region in the atmosphere where the temperature is optimal for the existence of life. It has been suggested that this region may be inhabited. I feel that the difficulties implicit in the evolution of life in a region far from a solid or liquid surface are so considerable that this is unlikely, but it is certainly not impossible. It is also just possible that life evolved at the surface long ago, when conditions were more favorable.

Similar arguments apply to the two major planets, Jupiter and Saturn. These planets are composed in the main of hydrogen and helium. They do not have solid surfaces, so if life evolved at all it must have been in the atmosphere. The temperature at the top of the atmosphere of Jupiter and Saturn are much lower than the one on Venus, but deep in the atmospheres there are believed to be regions at temperatures that are suitable for life. Again, the possibility that life has evolved in these regions cannot be dismissed entirely, but I think the probability is small.

This leaves Mars as the most promising planet on which to search for evidence of life. In the remainder of this chapter we shall survey what is known about surface conditions on Mars and the likelihood that life could exist there. This subject is particularly interesting at the moment, since in 1976 we shall have the opportunity to land scientific instruments on the surface of Mars to look for evidence of life.

It is believed that the red color of Mars is due to large quantities of red, iron-containing minerals present on its surface. Photographs taken by a Mariner spacecraft orbiting the planet show that the surface is cratered and looks very much like that of the moon in many places. However, because Mars is the site of intensive volcanic activity and also because it has an atmosphere, the surface of the planet shows much more evidence of active, geological processes. Ground-based observations have detected dust storms that sometimes obscure large areas of the planet's surface, and pictures taken by Mariner spacecraft have confirmed these observations in a spectacular way and revealed regions where craters have been filled in rapidly, presumably by drifting dust.

The temperatures on Mars are more extreme than are those on earth; but, certainly, they are not such as to preclude the existence of earth-like life. At the equator, temperatures as high as 32°C (90°F) are recorded in the summer, but the temperature falls to −100°C at the coldest period of the Martian night. Temperatures at the poles range from −10°C to −130°C.

The pressure of the atmosphere at the Martian surface is only about 1% of that on the earth. The very thin atmosphere is composed very largely of carbon dioxide. In addition it contains a small amount of carbon monoxide. It is not surprising that there is only very little oxygen in the atmosphere, but the recent finding that nitrogen is absent came as a great surprise. The Martian atmosphere does contain some water, but very little.

The white polar caps are very prominent in winter. They appear, to the eye, as though they were made of snow or ice.

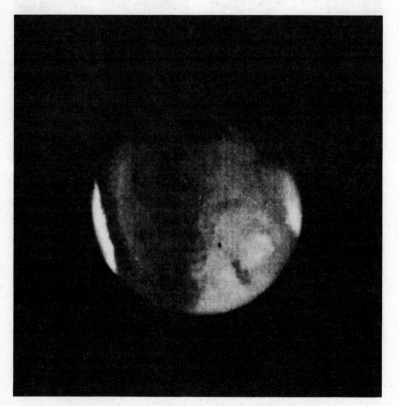

Figure 15.3. Mars as photographed by the Mariner VII Spacecraft at a distance of 534,398 miles from the Martian surface. The south polar cap is clearly visible at the bottom of the photograph. (Reproduced with permission of the National Aeronautics and Space Administration.)

Apparently, this is not the case. The main component of the Martian ice cap is known to be solid carbon dioxide. Water is a minor but important constituent.

The most intriguing changes that occur on the Martian surface are associated with the recession of the ice caps in summer. Each year, as the ice caps vaporize, a "wave of darkening" is initiated in the polar regions and spreads in

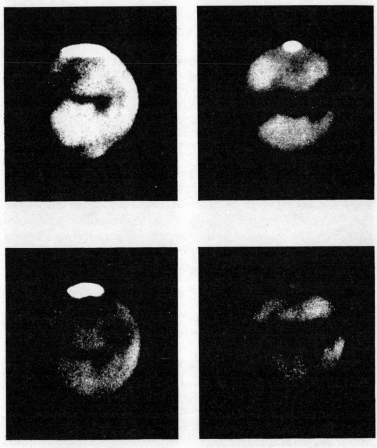

Figure 15.4. The "wave of darkening" that spreads towards the equator as the "ice-cap" melts. (Reproduced with permission from *Stars, Planets and Life* by Robert Jastrow, William Heinemann Ltd., London, 1967, p. 103.)

the direction of the equator. It has been suggested that the wave of darkening is caused by the seasonal growth of vegetation. However, there are a number of less exciting possibilities: the darkening could be caused, for example, by drifting dust.

Recent photographs of Mars taken from Mariner spacecraft, although of excellent quality, are not adequate to establish or rule out the presence of living organisms. Equivalent photographs of the Earth would not reveal unambiguous evidence of terrestrial life. Thus, there is no direct experimental evidence for or against the existence of life on Mars. The most recent indirect evidence shows that the atmosphere on Mars differs more from our atmosphere than was thought previously. This perhaps makes it slightly less likely that earth-like forms of life are present than seemed probable a few years ago. However, the odds have not changed very much.

Three characteristics of the Martian atmosphere seem at first sight to make it unsuitable for the evolution of life. It contains little water and no detectable nitrogen. In addition the atmosphere is not sufficiently reducing to make possible prebiotic syntheses of the kind discussed in Chapter 7. However, none of these observations argues strongly against the existence of life on Mars.

We must distinguish between conditions that permit the evolution of life and those that permit the persistence of life once it has evolved. It is doubtful that there is enough free water on Mars at present to permit the evolution of life to occur. However, it is likely that liquid water was once abundant, for the most recent Mariner pictures reveal surface features which are hard to interpret as anything other than river beds of comparatively recent origin. If so, life may have evolved on Mars in a wet period and gone underground (hibernated?). Recently a great deal of chemically bound water has been detected on minerals at the Martian surface, and there is no obvious reason why living organisms should not use it.

If there is no nitrogen on Mars, it is unlikely that earth-like forms of life can exist. However, Martian organisms may be able to survive in an environment containing much less

Figure 15.5 A 700 kilometer long sinuous valley on the surface of Mars. This valley may just possibly have been produced by the collapse of the roof over a surface lava flow. On the other hand it could have been formed by erosion—if so, water must once have been abundant on the Martian surface. (Reproduced with the permission of the National Aeronautics and Space Administration.)

nitrogen than is present on the earth. We cannot exclude the possibility that there is a percent or so of nitrogen in the Martian atmosphere and that this is enough to supply the needs of Martian organisms. Alternatively, some nitrogen may be present in the soil as nitrate, a form of nitrogen which could probably be utilized more readily than could molecular nitrogen.

The low abundance of reducing compounds in the Martian atmosphere was thought until recently to prove that the synthesis of organic molecules could not occur there. However, recent experiments have shown that this conclusion is incorrect. An atmosphere of the same composition as the Martian atmosphere was irradiated with ultraviolet light in the presence of a number of inert powders. In every case, the carbon monoxide in the atmosphere was converted into organic compounds; formaldehyde, a potential precursor of sugars, was the most abundant product.

In the past, the atmosphere of Mars may have been more reducing. In that case a wider range of organic compounds would have been available. If life evolved under such conditions, there is no reason why it should not have adapted itself as the atmosphere became more oxidizing. We know that just such an adaptation has occurred on the earth.

Many of the outstanding questions about Mars should be settled in 1976 when two Viking spacecraft are expected to land instrument packages on the surface. One group of instruments is designed to detect Martian microorganisms, if there are any in the neighborhood of the spacecraft. Another one will carry out a complete analysis of the organic constituents of the Martian surface. If no life is present, we should get an overall picture of the prebiotic organic chemistry of Mars. If we are fortunate enough to discover living organisms, a new perspective in biology will open up for us. I believe that the Viking Project is the most important scientific venture in the entire Space Program.

Intelligence in the Universe

16

In this chapter we shall consider the possibility that there are intelligent extraterrestrial societies, and the chance that, if they exist, we shall be able to communicate with them. We shall talk about earth-like forms of life only, because we know nothing about any other kind. This subject is well worth discussing, even if everything that can be said is highly speculative.

The probability that technological societies based on earth-like life exist outside the solar system can be discussed only after the problem has been broken down into a number of parts, namely:

1. How many planets (N) are there in the Universe?
2. What proportion of them (F_L) could support life?
3. What is the probability (F_p) that life evolves on a planet that can support life?

4. What is the probability (F_T) that a technological society develops on an inhabited planet?
5. How long, on the average, do technological societies survive?

If the answers to these questions were known it would be possible to use very elementary arguments to estimate the number of technological societies in the universe. Let us start with a very simple problem that illustrates the principle that is involved. Suppose that there are 1000 islands in a certain region of the Pacific Ocean, and that one tenth of them are inhabitable. Then it is obvious that there are about $1000 \times 1/10 = 100$ inhabitable islands. In exactly the same way if there are N planets in the universe and a fraction F_2 are of a type that can support life, then there are about $N \times F_L$ planets able to support life. The same type of argument can be used to show that $N \times F_L \times F_p$ planets not only can support life, but actually do become inhabited, and that $N \times F_L \times F_p \times F_T$ planets sooner or later come to harbor technological civilizations.

If we are interested in communication with other planets, we need to know what proportion would be occupied by intelligent civilizations at the moment of contact. Clearly this would be less than the proportion of planets that are occupied at one time or another in their history since we might attempt to make contact with the planet before an intelligent society had become established or after it had become extinct. Again the example of islands in the Pacific is instructive. Suppose that the supply of foodstuff on an island is adequate for only one year, but that it takes ten years to replace the supply once it has been used up. Then the chance of finding anyone on an inhabitable island on the occasion of a random visit would be at most one in ten. In the same way if an average planet survives for 10,000,000,000 years, but is usually made permanently uninhabitable within 10,000 years by technological development, then the chance of finding a resident technological society at the time of first contact would be 10,000/10,000,000,000 that is only one in a million. More generally the number of technological societies in the universe at any time is likely to be

$N \times F_L \times F_p \times F_T \times L_t/L_p,$* where L_t is the average lifetime of a technological society and L_p is the average lifetime of a planet.

Stars occur in widely separated clusters known as galaxies. Our own galaxy contains about 10^{11} stars; the number of stars in the universe is about 10^{20}. Unfortunately, even the nearest stars are so far away that astronomers would be unable to detect planets as small as the earth directly with the telescopes that are available today. Estimates of the number of planets in the universe must be based on indirect evidence.

A number of theories of star formation have been developed recently. They lead to the conclusion that planets no larger than the solar planets and also heavier planet-like objects (dark companions) are common. To some extent these predictions have been verified experimentally. Observations on nearby stars have shown that a surprisingly large proportion do have companions heavier than the largest solar planets. Since the large planets have been found, it is reasonable to believe that the small planets predicted by the theory are also there.

Another argument for the existence of a large number of planets is provided by studies of the rotational (spinning) motion of stars. The material in the interstellar dust clouds, from which stars are formed, rotates about the center of the cloud. A well-known law of physics, the law of the conservation of angular momentum, shows that this rotational motion cannot stop when the dust cloud condenses into a star. However, direct observation of stars similar to the sun shows, that they rotate very slowly.

A number of mechanisms have been proposed to account for the missing rotational motion. In the most plausible theory, it is suggested that most of the rotational motion has been transferred from the central star to a group of earth-like planets. Since we know that the planets possess most of the rotational motion in the solar system, this explanation is particularly attractive. However, we cannot be sure that it is correct; some other theories account for the loss of angular momentum without involving planets at all.

* This is a simplification of a formula given by F. Drake.

NGC 4565   Viewed edge-on

NGC 4216   Tilted 15 degrees

NGC 7331   Tilted 30 degrees

Figure 16.1. Some typical galaxies seen at various orientations. (Reproduced with permission from *Stars, Planets and Life* by Robert Jastrow, William Heinemann, Ltd., London, 1967, p. 27.)

Most astronomers agree that planets are likely to be very numerous. There could well be as many as 10^{10} in our own galaxy and 10^{19} in the universe. How many of them could support earth-like life?

Since life similar to that on earth could exist only in the presence of liquid water, earth-like life is improbable except in regions where the temperatures are moderate. Astronomers have tried to estimate the proportion of planets that have maintained a suitable range of temperature for a sufficiently long time to permit the evolution of life. They have concluded that such planets are quite common; there could be as many as 10^9 in our galaxy, and a correspondingly larger number in the universe.

Next we must estimate the probability that life does evolve, given a suitable enviroment. This is an extremely controversial topic. Some scientists believe that life inevitably occurs where conditions are favorable, but others believe that the evolution of life is extremely unlikely even under the best conditions. I believe it to be impossible to make a meaningful estimate, at the present time.

The evidence presented in Chapters 7 and 8 makes it probable that a rich prebiotic soup accumulates on a considerable proportion of earth-like planets. It is much harder to estimate the probability that living things evolve in such a soup. The evidence available at present does not indicate whether the evolution of biological systems is likely or not. We will not be able to decide until we know much more about the first steps in the evolution of life: the replication of polymers under prebiotic conditions.

Many writers have taken a more optimistic view of this problem, and some have claimed that life is almost certain to evolve in a suitable enviroment. They believe that, since life has evolved on earth, it is almost sure to have evolved on other earth-like planets. However, this seems to be an incorrect conclusion, based on a misunderstanding of probability theory.

The remaining questions that must be answered before we can estimate the probable number of technological societies in the universe are difficult to answer for similar reasons. So far, only one animal has developed sufficient intelligence to construct a technological society on earth. It is

Intelligence in the Universe

not possible, from this single example, to estimate how likely it is that intelligence would develop on some other inhabited planet, or whether a really intelligent species would bother to develop a highly sophisticated technology.

Estimates of the time for which technological societies are likely to survive depend more on the temperament of the writer than on relevant information. Optimists believe that the future of mankind is unlimited; pessimists believe that our chance of surviving for a further hundred years is poor. It is clear that, during the last hundred years, technological progress on the earth has been traumatic. We do not know whether this is a matter of historical accident, or whether the potentially self-destructive powers of technology always present a serious problem to advanced societies. Clearly, we are not in a position to estimate the average lifetime of a technological society.

It may be concluded that we cannot make meaningful estimates of some of the quantities that appear in our formula for the number of intelligent societies in the universe. There may be many of them or we may be alone. At an early stage in the development of a science, ignorance of this kind is not unusual and is certainly no cause for embarrassment. The problem of the "plurality of inhabited worlds" is part of our intellectual heritage. It is exciting to think that we are at last beginning to understand the chemical and biological principles that are involved in the evolution of life, and to build the telescopes with which to communicate with extraterrestrial civilizations, if they exist.

Communication with Extraterrestrial Civilizations

Even if extraterrestrial civilizations do exist, it may prove difficult to communicate with them if they are very far away. To appreciate this point it is necessary to have some feeling for the enormous distances that are involved. Light travels at the speed of 186,000 miles per second, or 5,880,000,000,000 miles per year. The distance even to the nearest stars is measured in light years, that is in units of 5,880,000,000,000 miles.

Our own galaxy is a flattened disc containing about 10^{11} (a hundred billion) stars; it is about 100,000 light years across and 2,000 light years thick. The galaxy closest to our own is about two million light years away; astronomical objects more than five billion light years away have been detected.

No radio signal or solid object can travel faster than the speed of light. This makes it difficult to communicate with distant planets. It is impossible to establish two-way communication with any planet more than 35 light years away in less than 70 years, for example. Only a few hundred stars fall within 35 light years of the earth, so only a tiny proportion of planets are close enough to be contacted in a human lifetime.

When objects are accelerated to speeds approaching that of light, their estimates of time are likely to be different from our own. A traveller on a spaceship that accelerated to almost the velocity of light, slowed down, and then returned home, would find that a longer time had elapsed on earth than on his spaceship. He might return as a young man, to find all his contemporaries had been dead for centuries.* In a similar way, an astronaut on an accelerating spaceship could travel very long distances in times that appeared quite short to him, although they would appear much longer to us. None of these paradoxes affect the conclusion that we cannot receive a reply from a distant planet in less than twice the time that light takes to travel there.

We are not yet in a position to travel to other solar systems, but in time this should be possible. Sooner or later we shall be able to make return journeys to the few hundred stars within 30 light years of the earth. Unfortunately, such journeys will probably not take place in our lifetimes. For future generations, space travel may provide the most direct form of two-way communication with extraterrestrial civilizations, if there are any close enough.

Two-way communication using radio or light waves is

* Most theoretical physicists believe that this is a legitimate deduction from the special theory of relativity, and likely to be true.

Intelligence in the Universe

restricted in the same way by the finite speed of light. If we sent out a message to one of the most distant stars in our own galaxy, it would take at least 100,000 years to get a reply. Two-way communication with a civilization in another galaxy would certainly take at least 4 million years. Even the prospect of conversing with the nearest stars, a few light years away, is daunting.

The prospects of picking up a message sent out spontaneously from another society are much better. We could, in principle, receive messages from any part of the universe—a message (or a spaceship) travelling at almost the velocity of light that was sent out from the nearest galaxy two million years ago, for example, should reach us about now. Of course, the individuals who sent the message would probably have died long ago.

Our ability to detect signals from other planets would depend on three factors. The distance away of the sender, the power of his signal, and the sharpness with which he beamed his message to the earth. A great deal of thought has been given to the problem of detecting weak signals from outside the solar system, and certain general conclusions have been reached.

The radiotelescopes that are used at the present time are not sensitive enough to permit us to eavesdrop on conversations between civilizations whose power consumption is similar to our own. Eavesdropping would be possible only if the signals that are used on other planets are much more powerful than any that can be generated on earth.

The detection of beamed signals would be much simpler. If any extraterrestrial civilization has set up a beacon to make its presence known, we might well be able to detect it. The first meeting of CETI, a society whose objective it is to communicate with extraterrestrial intelligence, was held at the Byurakan Observatory in Soviet Armenia during September, 1971. A group of American, European, and Soviet scientists, including two sceptical Nobel Prize winners, agreed that a limited search for extraterrestrial signals is justified at the present time.

What would messages of extraterrestrial origin be like?

Almost surely the technology on a planet that sent such messages would be more advanced than our own. The senders, therefore, would be at least as familiar with mathematics and physics as we are. They would also realize that anyone able to receive their messages must belong to a society capable of building telescopes. It seems probable, therefore, that the messages would include simple texts dealing with mathematics and technology. A great deal of work has already been done on the deciphering of texts of this kind. We could probably understand them once they had been identified.

It is hard to know what other information extraterrestrial societies would send us. A fascinating but gloomy prognosis is given in Hoyle's novel *A for Andromeda*. Some optimistic writers believe that the senders would be benevolent missionaries, determined to improve the lot of mankind. It will be left to the reader to write his own science fiction, and to decide for himself whether it would be wiser to answer the first messages we receive or to keep very quiet.

It is impossible to estimate the probability of our receiving intelligent messages from space. It may be foolish to take such a possibility seriously, or it may be unimaginative to do otherwise. The detection of another intelligent society in the universe would be one of the most important and exciting events in our history; if I had control of a radiotelescope, I would be willing to gamble a few percent of the watching-time to search for evidence of extraterrestrial intelligence.

UNIVERSITY OF PITTSBURGH
BOOK CENTER
4000 Fifth Avenue
Pittsburgh, Pa. 15213

9/0   103   8001   10

221   1   4.25 MDS
                       2   .58#MDS
                              4.83 STL
                               .04 TAX
                              5.00 ATD
                               .13 CDU

19/73   1 CSH   4.87 TTL

f the Main Argument

The earth was formed from a cloud of dust and gases about four and a half billion years ago. The cloud was made up very largely of hydrogen and helium, but these gases, together with most other volatile material, escaped during the condensation process. Our atmosphere and oceans are derived from gases that were expelled from the interior of the earth at a later date.

The primitive atmosphere was reducing. It contained water, methane, carbon dioxide, and nitrogen; in addition, some ammonia and hydrogen may have been present. We know that the atmosphere was still reducing a billion and a half years later, by which time organisms similar to algae or bacteria had evolved on the surface of the earth. When we talk about the origins of life, we mean the series of processes which occurred in the atmosphere, oceans, and lakes of the primitive earth and led, more than three billion years ago, to the appearance of the first living organisms.

We have no geological record of the events that occurred so long ago. However, it seems very likely that the first step

229

in the origins of life occurred when very simple organic molecules were formed in the earth's reducing atmosphere by the action of lightning, ultraviolet energy from the sun, shock-waves, and so forth. These compounds were washed into oceans and lakes where they reacted to form a complex mixture of organic substances, including many that were to form part of the most primitive living organisms.

The mixture of chemical compounds that was formed on the primitive earth is referred to as the prebiotic soup. Prebiotic chemistry is concerned with laboratory experiments that simulate the processes believed to have been involved in the formation of the prebiotic soup. The most striking success achieved so far in the study of the origins of life is the demonstration that the components of living organisms are particularly abundant in the mixture of organic chemicals formed in prebiotic reactions. Most of the important components of modern organisms, for example, amino acids, sugars, and nucleic-acid bases, have been synthesized in the laboratory under conditions that could have prevailed on the primitive earth.

The next stage in the origins of life must have been the concentration or "thickening" of the prebiotic soup. Evaporation was almost surely important, but freezing and adsorption on the surface of rocks or colloidal particles may also have played a part. Once the prebiotic soup was sufficiently concentrated, polymers resembling proteins and nucleic acids could be formed. Not much success has been attained in simulating these polymerization reactions in the laboratory, but some progress has been made.

The major intellectual problem presented by the origins of life is concerned with the next stage, the evolution of biological organization. How did a complex self-replicating organism evolve from an unorganized mixture of polymeric molecules? Little experimental evidence is available, so one is forced to attempt a speculative reconstruction of this phase in the origins of life.

The key to the understanding of the evolution of biological organization is the theory of natural selection. Before the evolution of complicated self-replicating organisms, natural selection must have acted on something much simpler, probably on polymeric molecules resembling nucleic acids. It is believed that nucleic acid-like molecules were formed in the prebiotic soup and were able to reproduce without the help of enzymes. The theory of natural selection then shows that those molecules that could replicate fastest would have become dominant in the prebiotic soup.

As the competition became fiercer, the more successful families of self-replicating molecules must have "learned" to make use of small molecules in their environment to help them to replicate even faster. The most important of these adaptations involved the amino acids; ultimately a family of self-replicating nucleic acids evolved to the point where they could begin to control the synthesis of polypeptide sequences that had useful catalytic properties. This adaptation led ultimately to the evolution of protein synthesis and the genetic code.

The evolution of protein synthesis is not understood in detail. One of the great challenges of the problem of the origins of life is to demonstrate in the laboratory how polynucleotides, without the help of preformed enzymes, could have replicated and begun to control the synthesis of peptides with determined sequences. Once this has been done we shall be well on our way to understanding the origins of the first living cells.

Until recently, theories of the origin of the prebiotic soup were based entirely on laboratory experiments. Within the last few years dramatic developments in astronomy have changed this situation. A number of small organic molecules have been detected in interstellar dust clouds far from the earth. The most abundant of these include formaldehyde, hydrogen cyanide, and cyanoacetylene. These three molecules had been proposed previously as three of the most important precursors of the prebiotic soup. Thus radioastronomy has provided evidence supporting strongly our general ideas about prebiotic synthesis.

The discovery of many of the twenty naturally occurring amino acids in the Murchison meteorite, a meteorite that could not have been contaminated with terrestrial organic material, is equally important. In fact the mixture of amino acids in the meteorite matches almost exactly the mixture of amino acids formed by the action of an electric discharge on a reducing gas mixture. Clearly, amino acids are formed in large amounts somewhere in the solar system far from the earth.

These new findings suggest that the range of organic compounds that can be formed under prebiotic conditions is small. It is not unlikely, therefore, that there are other planets on which rich prebiotic soups have accumulated. We do not know how likely it is that life would evolve on such planets, but we can see no good reason why earth-like life should not exist elsewhere in the universe. It is also possible that totally alien forms of life have evolved on other planets, but we are not as yet in a position to say much about this possibility.

Studies of the origins of life have reached a particularly exciting point. We hope within a few decades to understand how life on earth evolved from a random mixture of organic compounds. It may take a little longer to create novel self-replicating systems, but it is likely that this will be achieved within, say, a hundred years. Once we understand more about the evolution of biological organization, we should be able to say something quantitative about the probability that life exists elsewhere in the universe. Perhaps we shall not have to wait so long—a systematic search for signs of intelligent life elsewhere in the universe is likely to begin in the near future. If we succeed in making contact with an extraterrestrial civilization, our view of man's place in the universe will certainly change.

Bibliography

Origin of Life

Oparin, A. I., *The Origin of Life on Earth,* Academic Press, Inc., New York, 1957. Some of the detailed argument has now been superceded, but a classic and still a very interesting book.

Oparin, A. I., *Life, Its Origin, Nature and Development,* Academic Press, Inc., New York, 1964. A more recent but in many ways less satisfactory treatment.

Kenyon, D. H., and G. Steinman, *Biochemical Predestination,* McGraw-Hill Book Co., New York, 1969. An excellent advanced text.

Miller, S., and L. E. Orgel, *The Origins of Life,* Prentice-Hall, Englewood Cliffs, New Jersey, 1973. A book at the advanced undergraduate level emphasizing chemical problems.

Bernal, J. D., *The Origins of Life,* World Publishing Co., New York, 1965. A dictionary on the origins of life. Some of the discussion is now out of date.

Cairns-Smith, A. G., *The Life Puzzle,* University of Toronto Press, Toronto, 1971. An unorthodox view of the origins of life.

Molecular Biology

Watson, J. D., *The Molecular Biology of the Gene,* 2nd Ed. W. A. Benjamin, New York, 1970. Perhaps the standard treatment of molecular biology at the graduate or advanced undergraduate level.

Monod, J., *Chance and Necessity,* Alfred A. Knopf, New York, 1971. A very elementry account of molecular biology that concentrates on philosophical implications.

Woese, C., *The Origins of the Genetic Code,* Harper and Row, New York, 1967. A detailed discussion of the genetic code and how it evolved.

Evolution

Darwin, C., *The Origin of Species,* The origin of species (sixth ed. 1872) and *The Descent of Man,* Modern Library, New York, also reprint of 1st ed. (1859). Philosophical Library, New York. The most important book in the literature of biology and still excellent reading.

Eiseley, L., *Darwin's Century,* Victor Gollancz, London, 1959. A detailed but always interesting account of the development of evolutionary theory.

Stebbings, G. L., *Processes of Organic Evolution,* Prentice-Hall, Englewood Cliffs, New Jersey, 1966.

The Fossil Record

Eicher, D. L., *Geological Time,* Prentice-Hall, Englewood Cliffs, New Jersey, 1968.

Schopf, J. W., "Precambrian fossils and evolutionary events prior to the origin of vascular plants," Biological Reviews *45,* 319 (1970). A review of recent work on the earliest microfossils.

Astronomy and Geophysics

Jastrow, R., and A. G. W. Cameron, *The Origin of the Solar System,* Academic Press, Inc., New York, 1963.

Rittman, A., *Volcanoes and their Activity,* Interscience, New York, 1962.

Extraterrestrial Life

Shklovskii, I. S., and C. Sagan, *Intelligent Life in the Universe,* Holden-Day, San Franciso, 1966. An exciting treatment of the origins of life and of intelligent societies.

Hoyle, F., *Of Men and Galaxies,* University of Washington Press, Seattle, Washington, 1964.

Science Fiction

Hoyle, F., *The Black Cloud,* a Signet Science Fiction paperback edition, The New American Library, New York, 1959. An outstandingly imaginative account of a "foreign" intelligence.

Hoyle, F., *Andromeda Breakthrough,* Harper & Row, New York, 1964.

INDEX

235

237
Index